The allometry of growth and reproduction

The allometry of growth and reproduction

MICHAEL J. REISS
Hills Road Sixth Form College, Cambridge

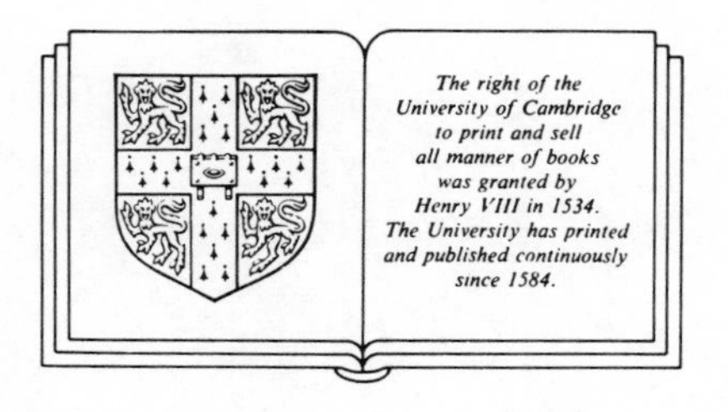

CAMBRIDGE UNIVERSITY PRESS
Cambridge
New York New Rochelle
Melbourne Sydney

Published by the Press Syndicate of the University of Cambridge
The Pitt Building, Trumpington Street, Cambridge CB2 1RP
32 East 57th Street, New York, NY 10022, USA
10 Stamford Road, Oakleigh, Melbourne 3166, Australia

First published 1989

Printed in Great Britain by the Alden Press, Oxford

British Library cataloguing in publication data

Reiss, Michael
The allometry of growth and reproduction.
1. Organisms. Life cycle. Physiological aspects
I. Title
574.3

Library of Congress cataloguing in publication data available

ISBN 0 521 360919

To Jenny

Contents

Preface xi

1 **Introduction** 1
The essentials of allometry 1
Materials and methods 4

2 **The scaling of average daily metabolic rate and energy intake** 7
Measurements of the scaling of ADMR on W 8
Reasons for the scaling of ADMR on W 11
Measurements of the scaling of energy intake on W 16
Reasons for the scaling of energy intake on W 20
Conclusions 21

3 **Why do larger species invest relatively less in their offspring?** 23
The model 23
Measurements for the interspecific scaling of E_{rep} on W 25
Alternative explanations for the interspecific scaling of E_{rep} on W 28
Conclusions 30

4 **The intraspecific relationship of parental investment to female body weight** 31
The model 31
Measurements for the intraspecific scaling of E_{rep} on W 32
Alternative explanations for the intraspecific scaling of E_{rep} on W 36
Conclusions 37

5 **Growth and productivity** 39
Growth curves 39

	Age at maturity, generation time and the duration of parental investment	43
	Energy budgets and growth	48
	Energy budgets and reproduction, and productivity/biomass ratios	54
	Why are there so many species?	62
	Home range sizes and *r* and *K* selection	64
	The cost of play	69
	Conclusions	71
6	**Quantitative models of body size**	75
	A review of existing models	75
	Sensitivities in body size models to changes in parameter values	88
	Conclusions	90
7	**Sexual dimorphism in body size**	91
	Male body weight and reproductive success	92
	A model of sexual dimorphism in body size	97
	Sexual dimorphism and the energy available for reproduction	111
	Conclusions	114
8	**Are larger species more dimorphic in body size?**	117
	A review of the evidence	118
	Why might larger species be more dimorphic?	122
	Conclusions	128
9	**Surface area/volume arguments in biology**	129
	Why cannot large animals rely on diffusion for gaseous exchange?	129
	Hypsodonty	134
	Other surface area/volume arguments that may be invalid	135
	When are surface area/volume arguments valid?	136
	Conclusions	137
10	**Prospectus**	139
	Allometry in plants	139
	Optimal organ size	140
	Allometry of parental investment	141
	Some difficulties in the allometric approach	142

Concluding discussion 143

Glossary of mathematical terms 145

References 151

Index 177

Preface

Organisms are adapted by natural selection to their physical and biological environments. Individuals differ within and between species in, among other things, such features as their body size, their age at first reproduction and the effort they put into growth and reproduction. The body of theory, generally known as life-history theory, that considers such features is extensive and growing rapidly (Stearns, 1980; Calder, 1984; Sibly & Calow, 1986). In a seminal review of these theories, and of the data that support them, Stearns (1976) concluded that we need more comprehensive theory that makes more readily falsifiable predictions. This view has been echoed by others (e.g. Peters, 1983). One way of viewing this book is to see it as presenting a number of life-history theories. I have tried to make the theories comprehensive, to make their assumptions explicit, to investigate the sensitivity of their conclusions to variations in the assumptions, and to test them whenever possible.

In common with most spheres of knowledge, the field of allometry has expanded very considerably within the last ten to fifteen years. It is perhaps no longer possible for a single text to provide a detailed review of all the work (see Peters, 1983; Calder, 1984; Schmidt-Nielsen, 1984). I hope this book has a single thread that runs through it, but I have tried to restrict myself to those areas where I feel I may have something new to contribute.

In essence, this whole book rests on two equations introduced in Chapter 2. These equations relate the scaling of average daily metabolic rate and energy assimilation intraspecifically to body weight, W. The major components of average daily metabolic rate scale at about $W^{0.5}$ to $W^{0.75}$ and the reasons for this are fairly well understood; accordingly, average daily metabolic rate is itself expected to scale on body weight as $W^{0.5-0.75}$, which it does. The intraspecific scaling of energy assimilation on body weight has been less extensively investigated, but there are *a priori* reasons to expect an exponent of around 0.67; energy assimilation does indeed appear to scale at abut $W^{0.67}$.

Having introduced these two fundamental equations, I then go on to see what conclusions can be drawn from them. The analysis in Chapter 3 relies on the deduction that the energy that adults can devote to reproduction equals their energy assimilation minus their non-reproductive energy requirements. As each of these scales on body weight with an exponent of less than 1, larger species are predicted to be able to invest relatively less in their offspring. This is indeed the case, Chapter 4 extends this model to consider the intraspecific scaling of the energy that females devote to reproduction. In species where adult female body weight changes little between successive reproductions, the intraspecific prediction is the same as the interspecific one, namely that the exponent relating the energy females devote to reproduction to body weight should be less than 1. In species where each female may reproduce over a wide range of body weights, however, the exponent is predicted to be greater. Analysis of data obtained from aphids, fish and isopods supports this prediction.

Chapter 5 is about growth and productivity. Almost none of the growth curves used in the literature has a physiological basis. A general equation for growth is suggested, based on the model in Chapter 4. A prediction is made for the intraspecific slope of log relative growth rate (percentage weight gained per day) on log body weight. This prediction is supported by using data from fish. Predictions are also made for the interspecific scaling of maximal growth rates on adult (female) body weight both for species where most growth is caused by parental investment, and for species where this is not the case. These predictions are again supported with data from reptiles, fish, birds and mammals. Predictions are also made for the interspecific scaling of age at maturity, generation time and duration of parental investment. These predictions are supported by data from viruses to giant sequoia trees. It is also possible to predict the proportion of an individual's energy budget that is devoted to growth as a function of an individual's size, and to predict the proportion of the total energy budget that is used for growth from conception to maturity. Light is also shed on the fact that productivity/biomass ratios scale at about $W^{-0.3}$. A major shortcoming of this chapter is that it fails to account for the very low productivity/assimilation ratios found in homeotherms. Contrary to popular belief, however, this is not a trivial consequence of the high metabolic rates of homeotherms; homeotherms also have much higher rates of energy assimilation than poikilotherms. Chapter 5 concludes with predictions for the relationship between the r and K values of a species, looks at the allometric scaling of home range size and considers the importance of allometric arguments to the questions 'Why are there so many species?' and 'Why is play mainly a feature of juveniles?'. A new explanation is proposed for the latter phenomenon.

Chapter 6 shows that almost all previous predictions of species' body weights are totally invalid. The purpose of the chapter is not, however, solely to condemn most previous work. A careful investigation into the form of the models used suggests why precise predictions of optimal body size are likely to be extremely difficult. Errors of 20% in the estimation of the most important parameters are shown to lead to 1000% errors in predictions of body weight! These problems are only exacerbated when the intraspecific dependence of lifespan on body size is considered.

In Chapter 7 the model first introduced in Chapter 3 and developed in Chapter 6 is extended to males to allow predictions of the degree of sexual dimorphism within a species. Data are first reviewed for the intraspecific scaling of male reproductive success on body weight and a general allometric form for the relationship is adopted. This relationship is then introduced into a general model of sexual dimorphism. It is concluded that although within a species females may be thousands of times heavier that males, males can never be more than about eight times heavier than females. It transpires that knowledge of just one variable – the exponent relating male reproductive success to body weight –

allows a precise quantitative prediction of the degree of sexual dimorphism within any species. It is also predicted that males can only afford to spend about 3–10% of their time trying to reproduce.

Two generalizations are frequently made about the degree of sexual dimorphism within a species. First, that dimorphism increases with the degree of polygyny; secondly, that larger species are more dimorphic. Chapter 8 first reviews the evidence for the association between size and sexual dimorphism, and then considers those theories that predict that larger species should show greater dimorphism in size. In several taxa larger species are more dimorphic, although there are exceptions. The main reason why a positive association between size and sexual dimorphism sometimes exists is probably because, on an evolutionary timescale, ecological factors such as food distribution affect both size and the opportunity for polygyny – polygynous species tend to be dimorphic – rather than because of a direct causal link between size and dimorphism. When the effects of polygyny on sexual dimorphism are removed, only in primates and small mammals is there still convincing evidence of a link between size and dimorphism. The theories that predict a causal link between size and dimorphism are generally unconvincing.

The universally accepted answer to the question 'Why do larger animals have lungs, gills or other specialized structures for gaseous exchange while smaller animals manage by diffusion?' is that the larger an organism, the smaller its surface area/volume ratio. Chapter 9 points out that this argument makes two implicit assumptions. First, that the oxygen requirements of an animal are proportional to its volume or mass. Secondly, that diffusion can only provide oxygen at a rate proportional to an animal's surface area. The former of these two assumptions is incorrect. The oxygen requirements of animals scale on mass with exponents of less than 1. Metabolic requirements are proportional to $W^{0.5\text{–}0.8}$ not $W^{1.0}$. Consequently, the surface area/volume argument does not explain why only small animals rely on diffusion to supply their oxygen needs. Consideration is also given to other instances of surface area/volume ratio arguments which may be incorrect. Despite all this, there are circumstances when surface area/volume arguments are valid.

The Prospectus, Chapter 10, mainly looks at areas of major ignorance to which allometric arguments might make a valuable contribution, but also considers weaknesses and difficulties with the allometric approach. Maynard Smith once wrote: 'Ecology is still a branch of science in which it is usually better to rely on the judgement of an experienced practitioner than on the predictions of a theorist. Theory has never played the role that it has in population genetics, perhaps because there is nothing in ecology comparable to Mendel's laws in genetics' (Maynard Smith, 1974). Allometry is a powerful tool, and while it would perhaps be an exaggeration to claim that its role in ecology will ever be as central as the role of Mendel's laws in genetics, the basic equations of allometry nevertheless share with Mendel's laws features of simplicity and generality. There is, I suspect, much that allometry has still to contribute to biology.

I am indebted to a large number of people, without whose essential help this book would never have been written. A substantial portion of the work included here was carried out at the Sub-Department of Animal Behaviour and the Large Animal Research Group in Cambridge. Especial thanks are due to Tim Clutton-Brock who supervised my Ph.D. and always encouraged me to think critically and originally. In addition, I owe particular thanks for their valuable thoughts and various kindnesses to Steve Albon, Mike Appleby, Pat Bateson, John Birks, Jenny Chapman, Rosemary Cockerill, Doug Easton, Francis Gilbert, Alan Grafen and Paul Martin. Jenny Chapman and Richard Sibly very kindly commented on the entire typescript, and R. McN. Alexander on Chapter 9. Steve Albon, Anthony Arak, Tim Clutton-Brock, Fiona Guinness, Georgina Mace and Brian Mitchell very generously allowed me to analyse unpublished data of theirs. I am also grateful to the Natural Environment Research Council and Trinity College, Cambridge for invaluable financial support.

this, there are circumstances [illegible] convergent [illegible] arguments are valid.

The Postscript (Chapter 9) [illegible] of [illegible] importance to which [illegible] in the [illegible] contribution, but also considers weaknesses and difficulties with the [illegible] approach. [illegible] still a branch of [illegible] in which it is [illegible] better [illegible] an expecting [illegible] predictions [illegible]. The [illegible] never [illegible] in [illegible] population [illegible] Mendel's [illegible] (Maynard Smith [illegible]) [illegible] would [illegible] in that [illegible] Mendel's [illegible] Mendel's laws [illegible] has still to contribute to biology.

I am indebted to a large number of people without whose support and help this book would be far [illegible]. The bulk of [illegible] portion [illegible] was carried out at the Department of Applied Biology and the Large Animal Research Group in Cambridge. Thanks are due to Tim Clutton-Brock who supervised my PhD and always encouraged me to think critically and [illegible]. I have particular thanks for [illegible] valuable thoughts and [illegible] to Steve Albon, [illegible], Patrick Bateson, [illegible], Innes Cuthill, Rosemary [illegible], [illegible], Sam [illegible] and [illegible] Martin. [illegible] Chapman and Richard Sibly [illegible] contributed [illegible] typing, and [illegible] Chapter 8 Steve Albon, [illegible], Tim Clutton-Brock, Fiona Guinness [illegible] allowed me to analyse unpublished data [illegible]. I am [illegible] grateful to the Natural Environment Research Council and [illegible] College, Cambridge for [illegible] financial support.

1

Introduction

The essentials of allometry

In its most general sense, allometry is any study of size and its consequences. In reality the power function

$$Y = \alpha X^{\beta} \tag{1.1}$$

where X and Y are size-related measures, and α and β are constants, is almost invariably used (Gould, 1966), and is in this book called the allometric equation. The almost universal adoption of Equation (1.1) in studies of size has been criticized, and in particular it has been argued that the linear equation

$$Y = mx + c$$

where m and c are constants, is sometimes more appropriate (D'Arcy Thompson, 1942; Smith, 1980). I will present functional arguments, in some instances, for and against the use of Equation (1.1). Where appropriate, statistical comparisons are made with other equations relating the two variables.

If X and Y are related to each other as in Equation (1.1), then by taking the logarithm of each side of the equation, we have

$$\log Y = \log \alpha + \beta \log X \tag{1.2}$$

A graph of $\log Y$ as a function of $\log X$ now produces a straight line with slope β. A graph of Y as a function of X produces a curve unless $\beta = 1$. If β does equal 1, the relationship between X and Y is said to be isometric. Isometry is a special case of allometry.

Figures 1.1 and 1.2 may help clarify Equations (1.1) and (1.2).

Figure 1.1 shows maximal rates of oxygen consumption for nine species of wild African mammals as a function of the mass of each species (Taylor *et al.*, 1980). (Maximal rates of oxygen consumption were measured while the animals were running flat out for about five minutes on a treadmill.) Figure 1.2 shows the same data but on a double-log plot, as in Equation (1.2). The points now fall on a straight line with gradient approximately equal to 0.8. The fact that a straight line can obviously be drawn through the points in Figure 1.2 means that the two variables are related allometrically by Equation (1.1) – in this case with $\beta \simeq 0.8$.

There are occasions when more complicated techniques of presenting and analysing allometrical relationships are required (Mosimann & James, 1979), but Equations (1.1) and (1.2) will suffice for this book.

Donhoffer (1986) maintained that the slopes of regressions calculated by using Equation (1.2) depend on the units used for the independent variable. This paper would, had its central argument been valid, have destroyed the value of almost every work using techniques of allometrical analysis. For example, Donhoffer would

Figure 1.1. Maximal rates of oxygen consumption ($\dot{V}_{O_2 max}$) for nine species of wild African mammals plotted as a function of body mass on linear coordinates. (Data from Taylor *et al.*, 1980.)

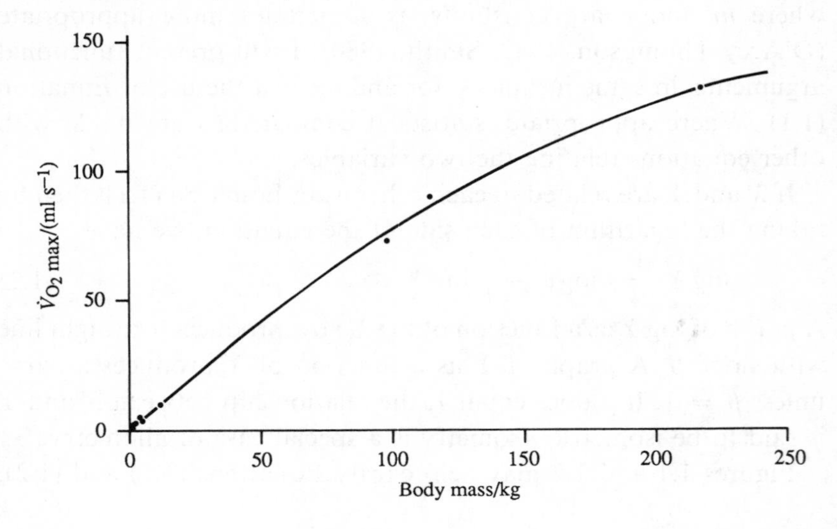

argue that the slope of the line in Figure 1.2 would have been different had grams rather than kilograms been used as the independent, X, variable. (Donhoffer's conclusion was never checked – rather it was arrived at implicitly from his valid demonstration that the percentage difference between, say, $W^{\frac{2}{3}}$ and $W^{\frac{3}{4}}$ depends on the units used for W.) It is easy, however, to show that the slopes of calculated regressions do not depend on the units used (Reiss, 1986c).

Suppose that we have the relationship

$$f(W) = aW^b$$

where a and b may or may not have been determined empirically.

We can substitute

$$W = k\tilde{W}$$

Figure 1.2. Maximal rates of oxygen consumption ($\dot{V}_{O_2 max}$) for the same nine species of wild African mammals as in Figure 1.1 plotted as a function of body mass on logarithmic coordinates.

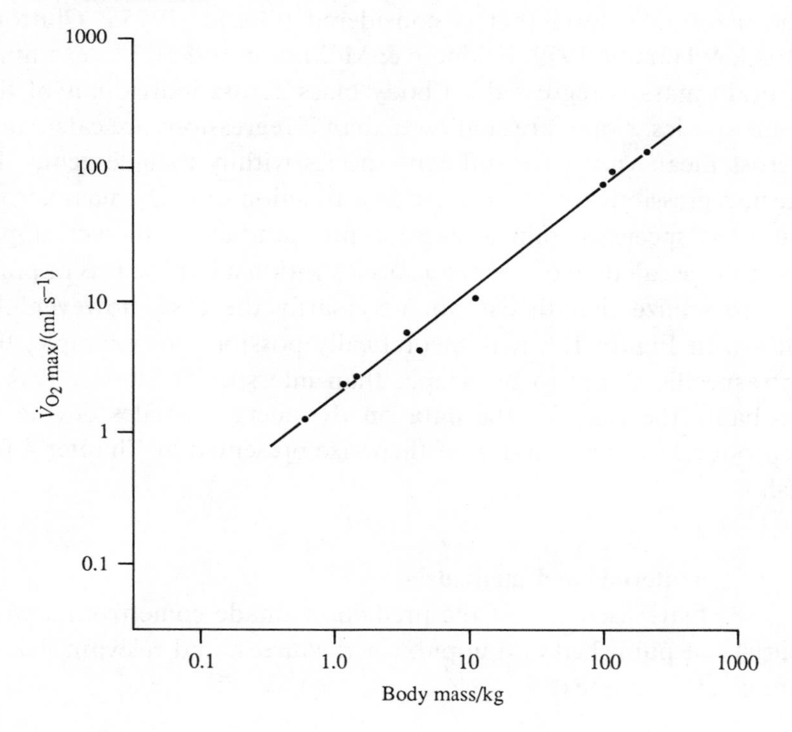

(e.g. $k = 1000$ to convert between W in grams and $\tilde{W}$ in kilograms).

We now have

$$f(W) = a(k\tilde{W})^b = ak^b\tilde{W}^b$$

so the slope, b, has not changed.

As an example, for cubes of the same density, length is proportional to $W^{\frac{1}{3}}$, where W is the mass of each cube, whatever the units of length or mass.

These conclusions remain unaltered if the points from which the regression is calculated do not all lie exactly on the calculated regression line.

Interspecific versus intraspecific allometry

It is often the case that size-dependent variables show progressively shallower slopes, when regressed on body size, the lower the taxonomic level that is considered (Gould, 1975a; Clutton-Brock & Harvey, 1979; Feldman & McMahon, 1983). For example, if brain mass is regressed on body mass across individuals of the same species, slopes are shallower than if regressions are calculated across mean values for different species within a single genus. In turn, regressions for brain mass as a function of body mass across different species within a single genus produce shallower slopes than those calculated across all species within a family. It is important to realize that this is not necessarily the case, however. As shown in Figure 1.3, it is theoretically possible, for example, for intraspecific slopes to be steeper than interspecific slopes. This is probably the case for the data on the energy females devote to reproduction as a function of their size presented in Chapter 4 for fish.

Materials and methods

Data used to test the predictions made come from a wide variety of published and unpublished sources, and relevant details are given in the text.

The main technique used is optimization; it is assumed that something maximizes something else. Three problems occur: first, to decide what or who is doing the optimizing; secondly, to decide what is being optimized; thirdly, to determine the constraints acting on the system. These are discussed in each chapter as appropriate. The problems of optimization in biology have been extensively discussed (Gould, 1978; Lewontin, 1978; Maynard Smith, 1978; Oster & Wilson, 1978; Clutton-Brock & Harvey, 1979; Gould & Lewontin, 1979; Maynard Smith, 1980; Krebs & McCleery, 1984; Krebs & Davies, 1987). In particular, Maynard Smith (1978) argued that a frequent difficulty is that conclusions depend on unstated assumptions. A particular effort has therefore been made here to make the assumptions explicit.

Probability tests are two-tailed. When a value for the strength of a relationship between two variables is needed, Spearman rank

Figure 1.3 Hypothetical relationship between a dependent variable and body mass showing that intraspecific relationships can theoretically have steeper slopes than interspecific ones.

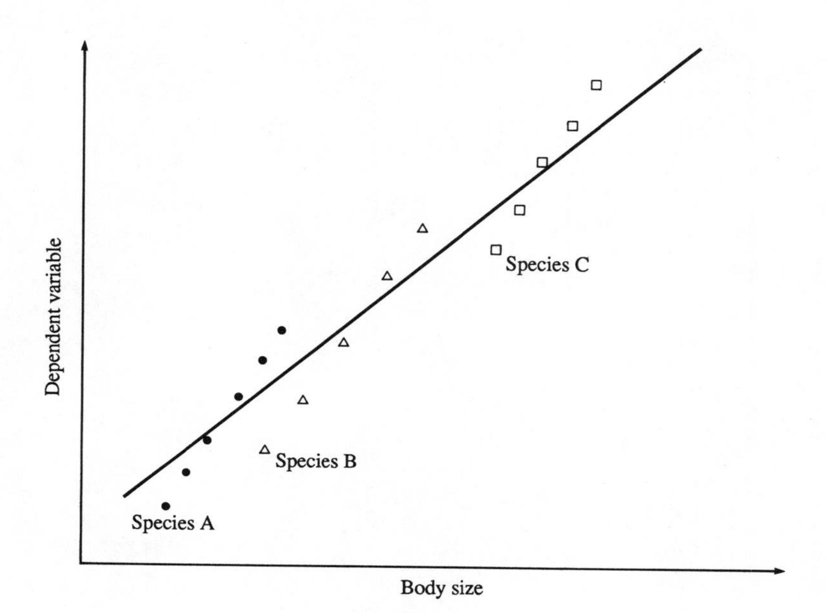

correlation coefficients, r_s, and Pearson product-moment correlation coefficients, r, are used as appropriate (Sokal & Rohlf, 1969). When the slope of a regression is required, least squares regression or reduced major axis regression have been used (cf. Kermack & Haldane, 1950; Kidwell & Chase, 1967; Zar, 1968; Glass, 1969; Brett & Glass, 1973; Ricker, 1973, 1975; Harvey & Mace, 1982). The debate on the most suitable regression technique periodically resurfaces in different areas of biology (e.g. Zar, 1968; Ricker, 1973; Harvey & Mace, 1982). Least squares regression assumes that only the y, dependent, variable is measured with error. Reduced major axis regression assumes that both variables are measured with equal error. Fortunately, for the high values of r that often occur with allometric plots, the results from the two methods are similar because they differ by a factor r: the slope calculated by reduced major axis regression equals the slope calculated by least squares regression divided by r. Both least squares and reduced major axis regression assume that the errors are normally distributed and of equal variance (Zar, 1968; Havlicek & Peterson, 1976). Unfortunately, it is not yet clear how robust regression analyses are to violations of these assumptions.

2

The scaling of average daily metabolic rate and energy intake

Average daily metabolic rate (ADMR) is the average metabolic rate of an active animal over a 24-hour period, at temperatures that the animal would normally experience in its natural environment, and when given access to food and water. Nagy (1987) provides a useful catalogue of the various terms that have been used to stand for realistic estimates or measurements of the daily energy expenditure of organisms in the field. These include, in addition to ADMR, 'field metabolic rate' (FMR), 'daily energy budget' (DEB) and 'daily energy expenditure' (DEE). ADMR is typically measured in millilitres of O_2 per hour. In this chapter it will be shown that intraspecifically, E_{req}, a free-living adult female's energy requirements per unit time for everything except reproduction, can be modelled as

$$E_{req} = k_1 W^a \qquad (2.1)$$

where k_1 and a are constants and W is body weight (strictly, body mass).

It will also be shown that intraspecifically, E_{in}, a free-living individual's energy intake (strictly, energy assimilation) per unit time, can be modelled as

$$E_{in} = k_2 W^b \qquad (2.2)$$

where k_2 and b are constants and W is again body weight.

On occasions, I will require the interspecific exponents relating E_{req} and E_{in} to body weight, so this chapter will also review data on these exponents.

Measurements of the scaling of ADMR on *W*

Intraspecific data on a

Table 2.1 lists intraspecific values of a and k_1 measured in mammals. Although the statistical methods used to determine a and k_1 are usually not stated, least squares regression is presumably the most common technique. Mace, Harvey & Clutton-Brock (1983) argued, however, that reduced major axis regression should be used. In reality, metabolic rate is probably less accurately measured than body weight. Consequently the real values of a probably lie somewhere between the values calculated from least squares and reduced major axis regressions. In almost none of the cases in Table 2.1 can confidence limits for either a or k_1 be calculated as insufficient data are reported in the original papers.

Grodziński (1971) forced the slope of ADMR on W to be 0.50 for the four species *Sorex cinereus*, *Clethrionomys rutilis*, *Glaucomys sabinus* and *Tamiasciurus hudsonicus*. Data for these species are nevertheless included in Table 2.1 as Grodziński wrote that 'the exponent of the weight regression is in all cases closer to 0.50 than to the well-known exponent of 0.75 for the basal metabolic rate'.

The data in Table 2.1 suggest that the intraspecific exponent relating average daily metabolic requirements to body weight usually lies between about 0.5 and 0.75. More data are available for mammals than for any other class of organisms. In birds, Mertens (1969) found that nestling great tit, *Parus major*, metabolism was proportional to $W^{0.67}$, whereas the values for the daily energy budgets of the 11 species for which sufficient data exist fall on approximately the same intra- and interspecific allometric curve (Figure 5.12 of Kendeigh, Dol'nik & Gavrilov, 1977). In the house sparrow, *Passer domesticus*, the intraspecific equation for ADMR on W from hatching to independence is given by

$$\text{ADMR} = 11.3W^{0.83}$$

and the corresponding equation for the black-bellied tree duck, *Dendrocygna autumnalis*, is

$$\text{ADMR} = 14.2W^{0.78}$$

where W is in grams and ADMR is in millilitres of O_2 per hour (Kendeigh *et al.*, 1977).

Table 2.1. *Intraspecific equations of average daily metabolic rate,* ADMR, *in millilitres of O_2 per hour, in relation to body weight, W, in grams, in mammals.* ADMR $= k_1 W^a$. *N is the number of points used to calculate the regression equation, r is the correlation coefficient of the regression, and where possible standard errors of a (± 1 SE) are given*

Species	±1 SE	a	k_1	r	Weight range	N	Source
Microtus pennsylvanicus		0.52	20.6		15–45 g	26	(1)
Microtus arvalis		0.52	19.5		12–42 g	45	(2)
Microtus agrestis		0.47	19.5		13–29 g	21	(3)
Microtus agrestis	0.09	0.69			11–30 g	49	(4)
Microtus oeconomus		0.44	27.4		17–38 g		(5)
Sorex cinereus		0.50	32.5		3–5 g		(5)
Clethrionomys rutilus		0.50	22.0		19–33 g		(5)
Glaucomys sabrinus		0.50	25.0		161–189 g		(5)
Tamiasciurus hudsonicus		0.50	29.6		231–275 g		(5)
Micromys minutus		0.47	22.8		5–15 g	19	(6)
Arvicola terrestris		0.43	25.5		39–123 g	44	(7)
Mustela nivalis	0.11–0.30	0.53–0.84	6.41–50.5	0.91–0.99	64–193 g	4–6	(8)
Cricetus ericetus		0.12	162.9		187–629 g	66	(9)
Alouatta palliata		0.95	1.12		3200–9430 g	6	(10)

Sources: (1) Wiegert, 1961; (2) Grodziński, 1967; (3) Hansson & Grodziński, 1970; (4) Ferns, 1979; (5) Grodziński, 1971; (6) Górecki, 1971; (7) Dorżdż, Górecki, Grodziński & Pelikán, 1971; (8) Moors, 1977; (9) Górecki, 1977; (10) Nagy & Milton, 1979.

Intraspecific measurements of the dependence of metabolic rate on body weight have also been made for poikilotherms. The most extensive data come from fish. Active metabolism (comparable to ADMR) scales at $W^{0.99}$ in sockeye salmon, *Oncorhynchus nerka*, with 95% confidence limits of the exponent equal to ± 0.01 (Brett & Glass, 1973), and as $W^{0.81}$ in rainbow trout, *Salmo gairdneri*, with 95% confidence limits of the exponent equal to ± 0.08 (Staples & Nomura, 1976). Standard metabolism – where no movement occurs – typically scales at about $W^{0.80}$, with a range from $W^{0.5}$ to $W^{1.0}$ (Glass, 1969; Fry, 1971; Brett & Groves, 1979). Intraspecific measurements for the value of *a* in Equation (2.1) have also been made for some invertebrates and are summarized in Table 2.2. The extent to which these studies investigated average daily metabolic rates rather than basal metabolic rates probably varied from study to study. In any case, the distinction between the two is less well-defined in poikilotherms, for which the concept of thermoneutrality does not apply.

The data from birds, fishes and invertebrates confirm the impression gained from mammals that intraspecific measurements of *a* lie beneath 1.0. Some of the invertebrate measurements of *a* exceed 1.0, but it is perhaps premature to state confidently that there are exceptions to the tendency for *a* to lie beneath 1.0. Intraspecific exponents sometimes vary depending on whether the database includes juvenile (growing) individuals or is restricted to adults of differing weights. Zeuthen (1953) noticed a tendency for the exponent of energy requirements on body weight to be greater in juveniles than in adults or near-adults, and this conclusion definitely holds for the more recent study by Epp & Lewis (1980) on three copepod species (see Table 2.2).

Interspecific data on a

The best interspecific data for the scaling of ADMR on *W* come from mammals and birds and are summarized in Table 2.3. It is clear that as is the case for the intraspecific exponent relating ADMR to *W*, the interspecific exponent lies significantly below 1.0. Insufficient data are available to see for sure whether this is so for poikilotherms, but it is difficult to believe that this is not also the case (see Bennett & Nagy, 1977; Nagy, 1982; Peters, 1983).

Reasons for the scaling of ADMR on W

The allometric dependence of average daily metabolic rate on body weight is understood to some extent. It used to be thought that basal metabolic rate was proportional to an animal's surface area, i.e. $W^{0.67}$ (reviewed by Kleiber, 1975). Several theories predict this (Sarrus & Rameaux, 1839; Maynard Smith, 1968; Wilkie, 1977). However, interspecifically, basal metabolic rate scales at about $W^{0.75}$ and not as $W^{0.67}$ (Kleiber, 1947). Why this is so is still a matter for debate. McMahon (1973, 1975) produced a model that predicts this relationship. His theory, however, assumes that the critical factor in the design of animals, plants and other organisms is that they are built according to the principle of elastic similarity. Essentially this means that organisms must be built strongly enough to stand under their own weight. It is difficult to see how the assumptions of McMahon's model are met by unicellular organisms, for example, yet in these too metabolic rate scales interspecifically as $W^{0.75}$ rather than $W^{0.67}$. Additionally, McMahon's theory predicts that bone lengths should scale as $W^{0.25}$ rather than as $W^{0.33}$, as predicted by geometric similarity. This appears to be the case only in the Bovidae. In other terrestrial mammals bone lengths (femur, tibia, metatarsal, humerus, ulna, metacarpal) scale interspecifically as $W^{0.35}$, with 95% confidence limits of the exponent equal to ± 0.03 (Alexander, Jayes, Maloiy & Wathuta, 1979). Blum (1977) pointed out that whereas the surface area of a three-dimensional sphere is proportional to its volume to the two-thirds power, the surface area of a four-dimensional sphere is proportional to its volume to the three-quarters power. Blum made various suggestions as to what this fourth dimension might be, and Boddington (1978) suggested that it might be time. If Blum and Boddington are correct, the interspecific scaling of basal metabolic rate to the three-quarters power of body weight follows from a four-dimensional surface law.

A different approach to explain the scaling of basal metabolic

Table 2.2. *Intraspecific exponents for metabolic rate on body weight in invertebrates. Symbols as in Table 2.1*

Species	a	±1 SE	r	Weight range	N	Source
Asellus aquaticus, woodlouse, 10 °C	0.79	0.07	0.76	4–60 g	104	(1)
15 °C	0.83	0.06	0.86	6–84 g	90	(1)
Branchinecta gigas, fairy shrimp, males	0.92			2–47 mg	26	(2)
females	0.88			0.3–86 mg	23	(2)
larvae	0.36			19–80 μg	11	(2)
Alaskozetes antarcticus, mite, 0 °C	0.77	0.04	0.92	4–255 μg	59	(3)
5 °C	0.69	0.04	0.91	3–238 μg	55	(3)
10 °C	0.71	0.02	0.97	2–259 μg	68	(3)
15 °C	0.97	0.10	0.85	20–234 μg	35	(3)
Idotea baltica, isopod, males	1.07		0.79	2–162 mg	95	(4)
females	1.20		0.70	3–57 mg	70	(4)
juveniles	1.11		0.98	71–2060 μg	28	(4)
Plectus palustris, nematode	0.75	0.02		71–2450 ng	108	(5)
Tetranychus cinnabarinus, mite	1.21		0.91	2–24 μg	49	(6)
Phytoseiulus persimilis, mite	0.97		0.80	3–23μg	61	(6)

Macrosiphum liriodendri, aphid, 15 °C	0.99		0.95	13–185 μg	24	(7)
20 °C	0.76		0.91	19–210 μg	32	(7)
25 °C	0.94		0.93	14–201 μg	40	(7)
Mesocyclops brasilianus, copepod, nauplii	1.08	0.10			30	(8)
copepodids & adults	0.56	0.04			88	(8)
Notodiaptomus venezolanus, copepod, nauplii	1.09	0.10			9	(8)
copepodids & adults	0.75	0.09			14	(8)
Thermocyclops hyalinus, copepod, nauplii	0.87	0.15			12	(8)
copepodids & adults	0.26	0.10			39	(8)
Cyclops bicuspidatus, copepod, 4 °C	0.30	0.04	0.89	82–9400 ng	50	(9)
6 °C	0.26	0.03	0.88	46–6300 ng	50	(9)
8 °C	0.51	0.07	0.87	75–8000 ng	50	(9)
10 °C	0.39	0.07	0.79	170–9000 ng	55	(9)
12 °C	0.36	0.06	0.79	63–9000 ng	55	(9)
Atta sexdens, ant	0.79			1.4–14.2 mg		(10)

Sources: (1) Prus, 1976; (2) Daborn, 1975; (3) Young, 1979; (4) Strong & Daborn, 1979; (5) Klekowski, Schiemer & Duncan, 1979; (6) Thurling, 1980; (7) Van Hook, Nielsen & Shugart, 1980; (8) Epp & Lewis, 1980; (9) Laybourn-Parry & Strachan, 1980; (10) Wilson, 1980.

rate has been used by Economos (1979a, b), who argues that the increase in gravitational load as size increases is responsible for the deviation of basal metabolic rate from the surface law. He maintains that basal metabolic rate is the sum of gravity's metabolic cost (which scales, he calculates, as $W^{0.89}$) and a surface-proportional metabolic expenditure independent of gravity (which scales as $W^{0.67}$). The sum of these two curves fits Kleiber's (1947) data slightly better than Kleiber's own curve.

More recently, some authors have questioned whether basal metabolic rate really does scale at $W^{0.75}$. Bartels (1982) plotted a graph of resting oxygen consumption of mammals versus body weight in a body weight range of 2.5 g up to 3.8 tons and found an exponent of 0.66 ($r = 0.98$). The exponent differed from Kleiber's because of data from very small mammals such as shrews. Across mammals weighing less than 260 g, the exponent was only 0.42 ($r = 0.76$). Across the larger mammals, the exponent was 0.76 ($r = 0.99$). It might be argued that the very notion of basal metabolism for such hyperactive animals as shrews is questionable. Heusner (1982) showed that within mice, *Peromyscus*, rats, cats, dogs, sheep and cattle basal metabolic rate scaled at $W^{0.67}$ (± 0.03). Feldman & McMahon (1983) agreed with Heusner's conclusion, but pointed out that the interspecific exponent was 0.752 ± 0.004. They suggested that one might expect the intraspecific and interspecific exponents to differ; the intraspecific exponent being 0.67 on the grounds of geometric similarity and the interspecific exponent being 0.75 on the grounds of elastic similarity.

For our purposes it is sufficient to accept that basal metabolisms do scale on body weights with exponents of 0.67–0.75, and that there are theories that predict this.

Basal metabolism is only one component of average daily metabolic rate that exceeds the former by a factor of about 1.5–3.0 (Gessaman, 1973; Moen, 1973; King, 1974; Mace, 1979; Nagy & Milton, 1979). Besides basal metabolism, the major component in homeotherms of average daily metabolic rate is heat loss. Thermal conductance, C, in the equation $Q_L = C(T_B - T_A)$, where Q_L is the heat loss, T_B is an animal's temperature and T_A is the ambient temperature, scales as $W^{0.51}$ in mammals, $W^{0.46}$–$W^{0.67}$ in birds, and

Table 2.3. *Interspecific and intergeneric equations of* ADMR, *in millilitres of* O_2 *per hour, in relation to body weight, W, in grams, in mammals and birds. Symbols as in Table 2.1*

Taxon	±1 SE	a	k_1	r	Weight range	N	Source
8 species of rodents		0.50	20.1		14–28 g	8	(1)
18 species of rodents	0.21	0.67	15.9	0.95	9–607 g	19	(2)
36 species of rodents		0.54	17.8		7–30 g	47	(3)
8 species of insectivores		0.43	29.6		3–21 g	15	(3)
22 genera of mammals		0.46	23.6	0.94	3–246 g	22	(4)
19 genera of rodents		0.77		0.89	3–212 g	19	(5)
13 species of marsupials	0.05	0.58		0.98	6–61 900 g	28	(6)
23 species of eutherians	0.02	0.81		0.98	13–84 000 g	46	(6)
12 species of birds	0.19	0.71	21.1	0.97		12	(2)
5 species of nectar-feeding birds		0.42	39.9	0.97		5	(7)
11 genera of birds		0.75		0.96		11	(5)
7 species of penguins	0.07	0.74		0.97		7	(8)
7 species of non-moulting penguins	0.07	0.68		0.91		7	(8)
13 species of moulting penguins	0.03	0.77		0.97		13	(8)
13 species of petrels	0.03	0.67		0.98		13	(8)
42 species of birds		0.61		0.99	3–25 200 g	42	(9)
96 species of birds	0.03	0.65		0.91	3–25 090 g	96	(10)
10 species of passerines	0.04	0.75		0.95	9–85 g	26	(6)
15 species of non-passerines	0.04	0.75		0.95	5–9440 g	24	(6)

Sources: (1) Grodziński & Górecki, 1967; (2) King, 1974; (3) French, Grant, Grodziński & Swift, 1976; (4) Mace, 1979; (5) Mace *et al.*, 1983; (6) Nagy, 1987; (7) MacMillen & Carpenter, 1977; (8) Croxall, 1982, pers. comm.; (9) Walsberg, 1983b; (10) Bennett & Harvey, 1987.

as $W^{0.37}$ in lizards (Herreid & Kessel, 1967; Lasiewski, Weathers & Bernstein, 1967; Calder, 1974; Kendeigh *et al.*, 1977; Bartholomew & Tucker, 1964; Mertens, 1969). Theory predicts that C should scale as $W^{0.50}$ or $W^{0.67}$ (Kleiber, 1972; Kendeigh *et al.*, 1977). It seems likely therefore that conductive heat loss is allometrically related to body weight, with an exponent of less than 1.0, as $(T_B - T_A)$ seems unlikely, intraspecifically, to depend strongly on body weight. The other components of heat loss – evaporation, convection and radiation – are approximately proportional to surface area (Monteith, 1973), and so scale at about $W^{0.67}$. Running costs per unit distance scale intraspecifically at $W^{0.72}$ (Fedak & Seeherman, 1979) although the reason for this is unclear (Alexander, 1977; Taylor, 1977). Similarly it costs larger animals less, relative to their body weight, to swim or fly a given distance and, again, the relationship is allometric (Schmidt-Nielsen, 1972). It could, however, be that intraspecifically larger individuals move further per day than smaller individuals – this is known interspecifically to be so (Clutton-Brock & Harvey, 1977; Mitani & Rodman, 1979) – which might make the relationship between the daily cost of locomotion and body weight less well-fitted by an allometric equation, or increase the exponent relating the two.

The major components of average daily metabolic rate seem therefore to be allometrically related to body weight, and to scale at about $W^{0.5}$–$W^{0.75}$. As the exponents are similar one would expect (Laird, 1965) average daily metabolic rate itself to scale at about $W^{0.5}$–$W^{0.75}$, as indeed it does (Tables 2.1, 2.2 and 2.3).

Measurements of the scaling of energy intake on W

Intraspecific data on b

The exponent b in Equation (2.2) relates energy assimilation to body weight. Van der Drift (1951) showed that in the millipede *Glomeris marginata* the ratio of food consumed to the two-thirds power of body weight was approximately constant for different-sized individuals. This ratio, subsequently called the Van der Drift constant, has also been found to have approximately the same intraspecific value in the millipede *Cylindrojulus silvanum* (Van der

Drift, 1951), three forest arthropods (Gere, 1956), eight other forest arthropods (Dunger, 1958) and the carnivorous arachnid *Mitopus morio* (Phillipson, 1960).

Direct measurements of the intraspecific exponents relating food consumption to body weight, as found in the literature, are listed in Table 2.4. Equation (2.2) strictly requires that energy assimilation rather than food intake be measured. Some of the data in Table 2.4 come from measurements of energy assimilation, and some from measurements of food intake. Despite the fact that different-sized animals, both within and across species, often differ in the quality of food that they ingest (Jarman, 1974; Clutton-Brock & Harvey, 1977; Mace, 1979; Cammen, 1980), both measurements fortunately seem to give similar values of b. In the fish *Megalops cyprinoides*, the exponent for food intake on body weight is 0.71 whereas the exponent for energy absorption on body weight is 0.70 (Pandian, 1967). In another fish, *Ophiocephalus striatus*, the corresponding figures are 0.76 and 0.77 (Pandian, 1967).

Interspecific data on b

Table 2.5 lists interspecific exponents relating energy intake to body weight. As with the intraspecific data, the difference between the slope of log energy assimilation on log body weight and log energy intake on log body weight appears to be small. Cammen's (1980) interspecific value for b equal to 0.74, as calculated by the slope of log food intake on log body weight, only changed to 0.77 when multiple regression was used to partial out the dependence on size of the quality of food ingested, despite a substantial decrease in food quality with increasing species size: the smallest species in his sample fed on foods of 7–93% organic matter, whereas the foods of the largest species were only 0.4–2% organic matter. Apparently size-dependent variation in assimilation efficiency (the proportion of the energy ingested that is absorbed or assimilated) is sufficiently small not substantially to affect values of b calculated by either method. Larger species also feed on poorer quality food in ungulates (Jarman, 1974; Clutton-

Table 2.4. *Intraspecific exponents, b, for food consumption on body weight. Other symbols as in Table 2.1*

Species	b	±1 SE	r	Weight range	N	Source
Megalops cyprinoides, fish	0.70			1–150 g	14	(1)
Ophiocephalus striatus, fish	0.77			2–124 g	14	(1)
Idotea baltica, crustacean	0.63	0.02	1.00	2–144 mg	11	(2)
Orchestia bottae, crustacean	0.67	0.04	0.97	10–73 mg	19	(2)
Menippe mercenaria, crustacean	0.67	0.06	0.91	37–539 g	24	(2)
Artemia salina, crustacean	0.69–0.79			100–5340 μg	4–5	(2)
Daphnia pulex, crustacean	0.64–0.89					(3)
Salmo trutta, fish	0.72–0.77	0.01–0.05				(4)
Idotea baltica, crustacean	0.63		0.78	0.1–86 mg	159	(5), (6)
Plectus palustris, nematode	0.77–0.88			72–2300 ng		(7)
Domestic cattle	0.61				293	(8)

Sources: (1) Pandian, 1967; (2) Sushchenya & Khmeleva, 1967; (3) Lampert, 1977; (4) Elliot, 1979; (5) Strong & Daborn, 1979; (6) Strong & Daborn, 1980; (7) Schiemer, Duncan & Klekowski, 1980; (8) Forbes, 1982.

Table 2.5. *Interspecific exponents relating food consumption to body weight. Symbols as in Table 2.1*

Taxon	Exponent	±1 SE	r	Weight range	N	Source
15 species of copepods	0.62	0.10	0.81	1–210 μg	24	(1)
10 species of crustaceans	0.80	0.02	0.98		114	(2)
11 species of forest floor arthropods	0.68	0.13		5–210 mg	11	(3)
19 species of birds	0.63	0.09			19	(4)
120 species of endothermic herbivores	0.72	0.02	0.97	4 g–1600 kg	120	(5)
150 species of endothermic carnivores	0.69	0.01	0.98		150	(5)
49 species of carnivorous reptiles and amphibians	0.82	0.03	0.97	580 mg–39 kg	49	(5)
19 species of benthic invertebrate deposit feeders and detritivores	0.74		0.97	0.2–2050 mg	19	(6)

Sources: (1) Ikeda, 1977; (2) Sushchenya & Khmeleva, 1967; (3) Reichle, 1968; (4) Calder, 1974; (5) Farlow, 1976; (6) Cammen, 1980.

Brock & Harvey, 1983), primates (Clutton-Brock & Harvey, 1977) and rodents (Mace, 1979).

Reasons for the scaling of energy intake on W

The scaling of energy assimilation on size is less well understood than the scaling of metabolic rate on size. The allometric nature of energy assimilation might be expected on the grounds that feeding is fundamentally a surface phenomenon (Gould, 1966). For example, if energy assimilation is limited by the transfer of nutrients across the gut wall, the exponent of energy assimilation on body weight should lie close to 0.67. It could, however, be argued that as digestive enzymes operate throughout an animal's gut, b should lie close to 1. It all depends on precisely what is limiting energy assimilation. In hummingbirds, for instance, it is not clear whether energy assimilation is limited by the rate of sucking up nectar, the rate at which flowers can be visited, the rate at which flowers replenish their nectar reserves or the rate at which the nutrients can be absorbed (Hainsworth, 1973; Brown, Calder & Kodric-Brown, 1978; Karasov, 1986).

Clutton-Brock & Harvey (1983) argue that in herbivores that select discrete food items that are smaller than their mouth size (such as the tips of leaves or shoots), bite size may be unrelated to body size; in herbivores that feed less selectively on swards whose length is less than that of the buccal cavity, bite size may be proportional to incisor breadth, which scales, as predicted by the criterion of geometric similarity, as $W^{0.33}$ among ruminants. Clutton-Brock & Harvey suggest that in most grazing species, bite size is probably seldom constrained by mouth volume, and predict that food intake may scale as, at most, $W^{0.33}$. I find it difficult to accept this; one would expect the length of food taken also to scale as $W^{0.33}$, leading overall to an exponent of 0.67 relating food intake to body weight. Some data for red deer are analysed in Chapter 6 and suggest a value for b nearer to 0.67 than to 0.33.

It will be shown in Chapter 4 that there are reasons to predict that intraspecifically a exceeds b. In Chapter 5 it will be suggested that interspecifically a and b have the same value.

Conclusions

Equations (2.1) and (2.2) are, like nearly all biological laws, only approximations. Nevertheless, the available data show that they may be close approximations. Indeed, very many anatomical and ecological measures are allometrically related to body weight (Brody, 1945; Calder, 1974, 1984; Apple & Korostyshevskiy, 1980). Occasionally, E_{req} or E_{in} are represented as depending on body weight by some equation other than the allometric one (e.g. Daborn, 1975). Such equations have the disadvantage that they lack any functional explanation. Additionally, there is no evidence that they fit the data better (*pace* Smith, 1980).

Both average daily metabolic rate and energy assimilation are allometrically dependent on body weight with exponents of between approximately 0.5 and 1.0. These conclusions hold both for intraspecific and interspecific relationships. The reasons for these dependencies are largely understood.

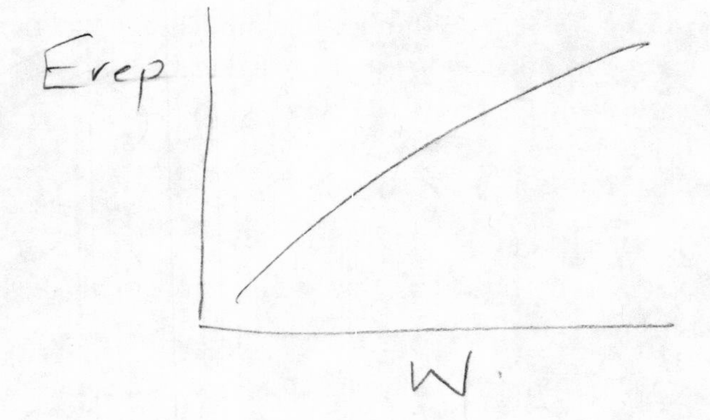
Erep
W

3

Why do larger species invest relatively less in their offspring?

It is known from several taxa (as critically reviewed below) that in larger species females invest less energy, relative to their body weight, in their offspring per unit time. The model presented here is meant to provide an explanation for this phenomenon. For convenience of expression, and because in the taxa from which data to test the model are available it is mostly true, it is assumed that females allocate their energy reserves for reproduction solely to their offspring. In general though, the model refers to the energy a female invests in other individuals, whether or not those offspring are her own offspring.

The model

Let E_{rep} be the energy an adult female can devote to reproduction per unit time. Then we have

$$E_{rep} = E_{in} - E_{req} \tag{3.1}$$

(as previously proposed by Ware (1980), Roff (1983) and others), where E_{in} and E_{req} were introduced in Chapter 2.

We can now substitute for E_{in} and E_{req} from Equations (2.1) and (2.2) into Equation (3.1) resulting in

$$E_{rep} = k_2 W^b - k_1 W^a \tag{3.2}$$

Figure 3.1 gives an example of how $k_1 W^a$, $k_2 W^b$ and E_{rep} depend on W intraspecifically. It is assumed that $1 > a > b$. Evidence that

a does indeed exceed b is deferred until Chapter 5. This is because the most convincing evidence that a does exceed b comes not from direct measurements of a and b, which were reviewed in Chapter 2, but from indirect measurements made either from individual growth curves or from the ontogenetic change in the proportion of an individual's energy budget that it devotes to growth.

It is clear from Equation (3.2) that we can predict that the exponent for an interspecific plot of E_{rep} as a function of W should be less than 1, and more specifically between about 0.5 and 1.0 (Reiss, 1985) as these are the exponents that interspecifically relate non-reproductive energy requirements and energy intake to body weight (Chapter 2).

It is important to note that the interspecific exponents relating energy assimilation and non-reproductive energy requirements to body weight are not necessarily the same as the intraspecific exponents. For example, a regression line plotted through a group of 10 g poikilothermic lizards and 20 g homeothermic mice would

Figure 3.1 Non-reproductive energy requirements, $k_1 W^a$, and energy intake, $k_2 W^b$, for adult females of different body weights. The curves are intraspecific curves (Equations (2.1) and (2.2)).

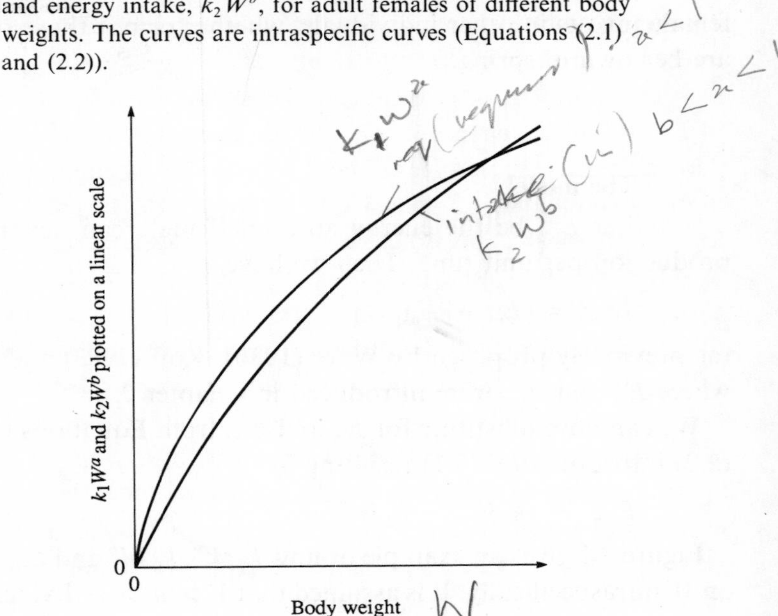

produce exponents in excess of 1. Similarly, an interspecific allometric plot of E_{rep} on W for these species would have a slope greater than 1. Obviously, the more uniform in ecology, physiology and feeding behaviour the species that compose an interspecific sample are, the closer the interspecific exponents for energy assimilation and non-reproductive energy requirements as functions of body weight will be to the corresponding intraspecific exponents. The intraspecific relationship of E_{rep} to W will be considered in Chapter 4.

Measurements for the interspecific scaling of E_{rep} on W

The energy females invest in their offspring is often difficult to measure (Hirshfield & Tinkle, 1975; Dittus, 1979). Furthermore, there has been a variety of indices of reproductive effort used to quantify parental investment (Calow, 1979; Tuomi, Hakala & Haukioja, 1983). Measurements of interspecific scalings of E_{rep} on W are listed in Table 3.1. Despite the variety of methods used to estimate E_{rep}, the relationship of E_{rep} to W is well described by the allometric equation and, as predicted, larger species do invest relatively less in their offspring per unit time.

Some comment is necessary, however, on the different measures of E_{rep} that have been used. E_{rep} is the energy invested in offspring per unit time, e.g. per year. There is probably a tendency for some of the exponents in Table 3.1 to overestimate the real slopes of $\log E_{rep}$ on $\log W$. This is because larger species generally reproduce less frequently. Consequently, measurements of clutch volume, clutch weight, litter weight and litter weaning weight, obtained on the occasions when individuals do reproduce, systematically overestimate E_{rep} in larger species. Clutch number, used by Petersen (1950) and Llewellyn & Brown (1985), has the disadvantage that larger species often produce larger offspring. Data on the interspecific scaling of offspring weight are not included in Table 3.1 as clutch and litter size vary systematically with adult body weight (Rahn, Paganelli & Ar, 1975; Blueweiss, Fox, Kudzma, Nakashima, Peters & Sams, 1978; Tuomi, 1980). In birds and mammals,

Table 3.1. *Interspecific exponents for E_{rep}, the energy females invest in reproduction, on W, adult female body weight. Standard errors of the exponents are given where possible. r is the correlation coefficient, and N is the number of points used in each regression*

Taxon	Measurement of E_{rep}	Exponent	± 1 SE	r	Weight range	N	Source
Spiders	Clutch number	0.84		0.94		28	(1)
Aphids	Clutch number	0.51–0.69		0.39–0.78	0.25–9.5 mg	29–83	(2)
Hoverflies	Clutch volume	0.95	0.10	0.89		30	(3)
Poikilotherms	Clutch volume	0.92		0.93	3 mg–140 kg		(4)
Salamanders	Clutch volume	0.64		0.90	140 mg–8900 g	74	(5)
Frogs	Clutch volume	0.90		0.98	0.7–120 g	23	(6)
Reptiles	Litter weight	0.88		0.98			(4)
Birds, Anatidae	Clutch weight	0.52	0.13	0.81	310 g–12 kg	149	(7), (8)
Birds, Phasianidae	Clutch weight	0.53	0.18	0.80	45 g–5 kg	50	(7), (8)
Birds	Clutch weight	0.74		0.92	4 g–100 kg		(3), (9)
Mammals	Litter weight	0.83			6 g–79 000 kg	114	(10)
Mammals	Litter weight	0.77	0.01	0.96	4 g–6000 kg	250	(11)
Mammals	Litter weaning weight	0.73	0.02	0.95	9 g–57 kg	100	(12)
Marsupials	Litter weaning weight	0.70		0.81		30	(13)
Mammals	Milk yield (kcal/day)	0.69	0.04		36 g–530 kg	20	(14)

Sources: (1) Petersen, 1950; (2) Llewellyn & Brown, 1985; (3) Gilbert, 1982; (4) Blueweiss *et al.*, 1978; (5) Kaplan & Salthe, 1979; (6) Crump, 1974; (7) Lack, 1968; (8) Rahn *et al.*, 1975; (9) Heinroth, 1922; (10) Leitch, Hytten & Billewicz, 1959; (11) Millar, 1981; (12) Millar, 1977; (13) Russell, 1982; (14) Hanwell & Peaker, 1977.

fecundity decreases with increasing species size, whereas the reverse occurs in poikilotherms. In birds and mammals, and to a lesser extent in some other taxa, parental investment continues beyond egg laying or birth. Clutch weight in birds is therefore bound to be an inadequate measure of E_{rep}. Unfortunately, in birds, the other components of E_{rep} are difficult to quantify (Ricklefs, 1974; Walsberg, 1983a). In mammals, weaning weight or milk yield are better indicators of E_{rep} than litter weight, although energetic investment can continue beyond weaning (Clark, 1978; Clutton-Brock & Albon, 1982), whereas peak daily milk yield clearly depends on the pattern and duration of lactation.

A further objection to the use of litter weaning weight as an index of E_{rep} is provided by Millar (1977) who argues that a better index is provided by $NW_w^{0.75}$ where N is litter size and W_w is offspring weight at weaning. This was said to be because $NW_w^{0.75}$ 'provides an estimate of the amount of energy that the offspring requires in relation to the non-breeding requirements of the female'. If Millar is correct, the exponents in Table 3.1 systematically overestimate the slope of $\log E_{rep}$ on $\log W$ not only for mammals, but also for other taxa. It can be shown, however, that litter weaning weight, NW, is a better index of E_{rep} than $NW_w^{0.75}$. Assume that a mother provides all the energy for her offspring until the termination of parental investment, when each offspring weighs W_w. For each offspring to increase in weight from W to $W + \delta W$, parental investment, PI, in a time δt approximately equals $Nk_2 W^b \delta t$. Now δW is proportional to $(k_2 W^b - k_1 W^a)\delta t$. Cumulative parental investment from conception to the end of parental investment is consequently given approximately by

$$\sum \mathrm{PI} \propto \int_0^{W_w} \frac{Nk_2 W^b}{k_2 W^b - k_1 W^a} \, \mathrm{d}W$$

Because $a \simeq b$, we have

$$\sum \mathrm{PI} \propto \int_0^{W_w} N \mathrm{d}W \propto NW_w \tag{3.3}$$

QED.

The only empirical data I know that relate to cumulative parental investment were collected, as one might have guessed, by Brody and are included in that remarkable text – Brody (1945). From observations on eight species of mammals and the chicken, Brody computed an exponent for cumulative parental investment on litter weight of 1.2, which is greater than the 1.0 I would predict from Equation (3.3).

E_{rep}, as I defined it, is the energy invested in reproduction per unit time. Larger species, however, live for longer. Lifespan is allometrically related to body weight with an exponent of about 0.15 (Blueweiss *et al.*, 1978) to 0.29 (Stahl, 1962). Combining this relationship with the data in Table 3.1 it is evident that over the course of their lives, the energy species devote to reproduction scales on body weight with exponents close to 1.0. One might expect this as lifetime energy intake and lifetime energy expenditure will also scale on body weight with exponents of approximately 1.0.

Alternative explanations for the interspecific scaling of E_{rep} on W

It might be thought that E_{rep} scales interspecifically below $W^{1.0}$ because there is an inverse interspecific and intraspecific relationship between adult survivorship and reproductive effort (Williams, 1966; Fleming, 1975; Ricklefs, 1977; Calow, 1979; Clutton-Brock, 1984), whereas larger species show greater adult survivorship (Stahl, 1962; Blueweiss *et al.*, 1978). However, if Equations (2.1) and (2.2) hold, and adult body size is constant on average from breeding event to breeding event, all of the available E_{rep} must be used in reproduction. An alternative answer is to realize that, for example, a 20 g lactating mouse produces each day approximately 35 times as much milk, weight for weight, as does a 2600 kg lactating elephant. It is impossible to account for such a disparity purely by the argument that larger species are more cautious in the amount of their available E_{rep} that they actually allocate to reproduction.

There have been previous attempts to explain why E_{rep} scales interspecifically below $W^{1.0}$. Huxley (1927), in attempting to explain why larger birds have relatively smaller eggs, suggested that the egg being an enormous cell, each successive increase in size would be achieved with proportionately greater difficulty. He also noted that nourishment for the growth of the egg must come through the egg's surface. Brody (1945) similarly pointed out that the visceral organs and surface areas that participate in the digestive, assimilatory, respiratory and secretory processes involved in milk production increase at a less rapid rate than the body as a whole. Lack (1968) argued that smaller bird species have proportionately heavier eggs because a smaller chick has a proportionately larger surface area, and so loses heat more rapidly, and so needs a larger food reserve at hatching. Blaxter (1971) noted that the exponent that related milk production in kilocalories per day to body weight was virtually the same as that which related metabolism to body weight, and simply stated that this was as expected. Crump (1974) thought that larger frogs have relatively small clutch volumes because 'larger species have a proportionately larger amount of supportive tissue'. Leutenegger (1976) maintained that in eutherian mammals maternal placental surfaces determine neonatal size, whereas Kaplan & Salthe (1979) argued that changes in distensibility of the body wall with changing size were the critical factor for salamanders.

These explanations may be correct in part, but by themselves they are unsatisfactory. In particular, they fail to explain why E_{rep} scales on W in similar ways in different taxa. Only by considering how both energy assimilation and non-reproductive energy requirements vary with body size can one explain how the energy available for reproduction varies with species size.

It would be nice to have some data for taxa like mayflies. In such semelparous taxa where adult females reproduce only once, and then only over a short period of time during which they feed little or not at all, E_{rep} might be expected to scale interspecifically as $W^{1.0}$ on the grounds that females would simply pour as much of themselves as possible into their offspring. It is difficult to see why smaller species should, in this situation, be able to devote a greater proportion of themselves to reproduction.

It may be that different groups of species have different interspecific values of the exponents relating non-reproductive energy requirements and energy assimilation to size. If this is the case, then some of the variation in the exponents relating E_{rep} to W listed in Table 3.1 may have a functional basis. For instance, it is possible that the energetic costs of hovering scale on body weight with an exponent of close to 1.0, although this is not known for certain (Hainsworth & Wolf, 1972a; Casey, 1981). If this is the case, it might be that in organisms such as hummingbirds and hoverflies, which spend a lot of time hovering and in which hovering is a major component of energy expenditure, a is close to 1.0. It is tempting to see confirmation of this in the fact that the largest exponent in Table 3.1 is indeed for hoverflies. However, it might be that larger species of hoverfly hover for less time each day, which would reduce the interspecific value of a. In birds, larger species spend a significantly smaller percentage of the active day in powered flight than do smaller species (Walsberg, 1983b). Data from Apodidae (swifts) would similarly be welcome especially as the percentages of time spent in flight in this family is presumably independent of species size.

Conclusions

Across species, analyses have shown that parental investment (measured as the number of calories an individual puts into its offspring per unit time) does not scale linearly with species weight. Larger species invest less, relative to their body weight, in their offspring per unit time. This trend is known to occur in several taxa. I suggest that the energy an adult female can invest in her offspring per unit time, E_{rep}, equals her energy assimilation minus her non-reproductive energy requirements. Each of these is allometrically related to body weight (Chapter 2) with interspecific exponents of between 0.5 and 1.0. It is predicted, therefore, that E_{rep} should scale interspecifically on body weight with exponents of between 0.5 and 1.0. The available data show that the energy females invest in their offspring per unit time scales interspecifically between $W^{0.52}$ and $W^{0.95}$.

4

The intraspecific relationship of parental investment to female body weight

Within most (perhaps all) species, larger females invest more in their offspring than smaller ones, either producing more offspring or investing more in each of them. This is hardly surprising. There is, however, no theory that predicts quantitatively just what the intraspecific dependence of parental investment on female body weight should be. In this chapter I first predict what the relationship between these two variables is, and then test the theory by using data from fish, aphids and isopods.

The model

In Chapter 3 the interspecific dependence of E_{rep} on W was considered. The intraspecific situation is more complicated because in some species individuals often reproduce before they reach their final body weight. Consider, first, however, species where females do not change much in weight during the reproductive phase of their lives. In this case Equation (3.1)

$$E_{rep} = E_{in} - E_{req}$$

is still valid and the intraspecific prediction for the dependence of E_{rep} on W is the same as the interspecific prediction: E_{rep} is predicted to scale on W with an exponent of between 0.5 and 1.0. For species, however, where each female reproduces several times over a wide range of body weights, Equation (3.1) is replaced by

$$E_{gro+rep} = E_{in} - E_{req} \qquad (4.1)$$

where $E_{gro+rep}$ is the energy available for growth and reproduction per unit time. In such species E_{rep} should increase more rapidly with W than as $W^{0.5}$–$W^{1.0}$. This is because data from the lower weights will come from individuals that are apportioning only some of their available energy for growth and reproduction to reproduction and the rest to growth (Reiss, 1987a).

Measurements for the intraspecific scaling of E_{rep} on W

The best data for organisms that change considerably in weight during the reproductive phase of their lives are reviewed by Wootton (1979) for fish. He compiled data from 124 studies of 62 species and expressed them as

$$F = aL^b$$

where F = fecundity, L = body length, and a and b are fitted constants (not related to the constants a and b used elsewhere in this book). Wootton found that b varied from 1.0 to 7.0 with a mean of 3.34. Assuming that body weight is proportional to the cube of body length (Ricker, 1973, 1979), E_{rep} scales intraspecifically, on average, as $W^{1.11}$ with 95% confidence limits of the exponent equal to ± 0.06. A more recent, although less extensive, review of 17 studies of 14 species of fish concluded that the exponent relating the weight of the gonads intraspecifically to body weight varied between 1.0 and 1.9 (Roff, 1983). Hence in fish the energy devoted to reproduction, as predicted above, scales intraspecifically on body weight allometrically with an exponent of greater than 1.0.

Ridley & Thompson (1979) reviewed data from eight studies of five species of *Asellus* (Crustacea; Isopoda). These do not reproduce several times over a wide range of body weights (Steel, 1961; Andersson, 1969; Ellis, 1971). Again, fecundity was expressed as an allometric function of body length. The exponent varied from 1.76 to 2.73 with a mean of 2.32. Assuming that body weight is proportional to the cube of body length, E_{rep} scales intraspecifically, on average, as $W^{0.77}$ with 95% confidence limits of the exponent equal to ± 0.09.

In Table 4.1 data on the intraspecific dependence of E_{rep} on female body weight are collated for aphids. The aim is to show the relationships between E_{rep} and W that have been published. Aphids were chosen, first, because parental investment ceases with the birth of an offspring and so fecundity provides a reasonable measure of E_{rep}, although in some aphid species heavier mothers also give birth to heavier offspring; secondly, because many authors have investigated how fecundity varies with size; and, thirdly, because there is considerable intraspecific variation in weight at reproduction. None of the papers suggests why E_{rep} scales on W as it does. Only Kempton, Lowe & Bintcliffe (1980) compare different methods of relating E_{rep} to W. They show that for *Myzus persicae* a negative exponential curve provides a significantly better fit than a straight line.

Aphids are similar to *Asellus* in that individuals vary little in weight during the reproductive phase of their lives. Consequently E_{rep} is predicted to be allometrically related to body weight with an exponent of between 0.5 and 1.0. When the two datasets from Kempton *et al.* (Figure 3 of their paper) are recalculated, linear regressions of log E_{rep} on log W have correlation coefficients of 0.98 in each case. The slopes are 0.47 with 95% confidence limits of ± 0.07, and 0.62 with 95% confidence limits of ± 0.08.

Of fish, isopods and aphids it appears that fish show the greatest range in exponents relating fecundity intraspecifically to body weight (from 0.33 to 2.33) – although data are available from more species of fish than from isopods and aphids. Wootton (1979) noted that short-lived species or those with poor post-spawning survival tend to have lower exponents than longer-lived species with good post-spawning survival. This is as one would predict from the model in this chapter. In species that are long-lived and have good post-spawning survival, it will be to a female's advantage to devote a high proportion of her energy requirements to growth when she is young and relatively small.

As noted in Chapter 3, larger species live for longer. The intraspecific relationship between size and longevity is less well known. If Boddington (1978) is correct in his presumption that the total metabolic expenditure by an animal is a constant independent of its

Table 4.1. *Intraspecific dependence of female reproductive effort on body weight in aphids: m, c and k are fitted constants*

Species	Dependent variable, E_{rep}	Independent variable	Relationship	Range of independent variable	Source
Acyrthosiphon pisum	Total fecundity	Body weight, W	$E_{rep} = mW + c$	5.0 from 575 to 2900 μg	(1)
Aphis fabae	Number of larvae	?	Larger individuals produce more larvae	data not given	(2)
Aphis fabae	Number of nymphs born in days 1–5, or 6–10, or 11–20	Body weight, W	$E_{rep} = mW + c$	5.3 from 184 to 981 μg	(3)
Aphis fabae	Number of nymphs	grid units [= area]	Larger individuals produce more nymphs	3.1 from 9.3 to 28.6 grid units	(4)
Aphis fabae	Number of well developed embryos in fundatrices	Body weight, W	$E_{rep} = mW + c$	1.4 from 1083 to 1496 μg	(5)
Aphis fabae	Number of well developed embryos in fundatrigeniae	Body weight, W	$E_{rep} = mW + c$	2.3 from 600 to 1352 μg	(5)
Aphis fabae	Number of well developed embryos in apterous exules	Body weight, W	$E_{rep} = mW + c$	3.5 from 355 to 1237 μg	(5)
Aphis fabae	Number of embryos	Body weight, W	$E_{rep} = mW + c$	4.0 from 328 to to 1296 μg	(6)
Aphis fabae	Number of embryos	Body weight, W	$E_{rep} = m \log W + c$	8.0 from 148 to 1188 μg	(6)
Brevicoryne brassicae	Number of progeny	Body weight, W	$E_{rep} = mW + c$	3.0 from 140 to 413 μg	(7)
Drepanosiphum platanoides	Number of young	Body weight, W	$E_{rep} = mW + c$	5.7 from 400 to 2288 μg	(8)
Eucallipterus tiliae	Number of nymphs	Body weight, W	$E_{rep} = mW + c$	5.1 from 144 to 730 μg	(9)
Metopolophium dirhodum	Total number of embryos, or number of embryos with pigmented eyes, or maximum daily fecundity, or number of nymphs in 20 days, or number of numphs in first 5 days, or number of nymphs in second 5 days, or number of nymphs in second 10 days	Body weight, W	$E_{rep} = mW + c$	5.5 from 273 to 1515 μg	(10)

Metopolophium dirhodum	Number of large embryos or total number of embryos	Body weight, W	$E_{rep} = mW + c$	Data not given	(11)
Myzus persicae	Number of large embryos	Body weight, W	$E_{rep} = m(1 - ce^{-kW})$	4.4 from 175 to 775 μg	(12)
Rhopalosiphum padi	Number of offspring produced on day 1, or 1–2, or 1–4	Body weight, W	$E_{rep} = mW + c$	5.3 from 150 to 800 μg	(13)
Rhopalosiphum padi	Number of large embryos, or total number of embryos	Body weight, W	$E_{rep} = mW + c$	Data not given	(11)
Rhopalosiphum padi	Potential fecundity (from dissection)	Body weight, W	$E_{rep} = mW + c$	3.0 from 204 to 604 μg	(14)
Sitobion avenae	Number of offspring	Body weight, W	$E_{rep} = mW + c$	3.6 from 280 to 1000 μg	(15)
Sitobion avenae	Total number of embryos, or number of embryos with pigmented eyes, or maximum daily fecundity, or number of nymphs in 20 days, or number of nymphs in first 5 days, or number of nymphs in second 5 days, or number of nymphs in second 10 days.	Body weight, W	$E_{rep} = mW + c$	9.4 from 182 to 1718 μg	(10)
Sitobion avenae	Number of large embryos, or total number of embryos	Body weight, W	$E_{rep} = mW + c$	Data not given	(11)

Sources: (1) Murdie, 1969; (2) Banks, 1964; (3) Dixon & Wratten, 1971; (4) Taylor, 1975; (5) Dixon & Dharma, 1980a; (6) Dixon & Dharma, 1980b; (7) Way, 1968; (8) Dixon, 1970; (9) Dixon, 1971; (10) Wratten, 1977; (11) Dewar, 1977; (12) Kempton *et al.*, 1980; (13) Dixon, 1976; (14) Wellings, Leather & Dixon, 1980; (15) Watt, 1979.

body weight, then the intraspecific relationship between longevity and size should be the same as the interspecific relationship. Size and longevity are positively correlated in male *Drosophila melanogaster* (Partridge & Farquhar, 1981), but negatively correlated in male mountain sheep (Geist, 1971).

One simplification of this chapter that may be important is the implicit assumption that only the difference between E_{in} and E_{req} can be used for reproduction. In contrast, semelparous animals like salmon and many plants (Harper, 1977) transfer nutrients and energy from their somatic tissues to their progeny before they die (cf. the discussion on mayflies in Chapter 3). In such species the slope of log E_{rep} on log W might be expected to equal or even exceed 1.0.

Alternative explanations for the intraspecific scaling of E_{rep} on W

There are few theories that predict the form of the intraspecific dependence of E_{rep} on W. Williams (1959) suggested for fish that fecundity might be limited by 'the mechanical restrictions on the amount of space in the body cavity and the necessity of maintaining locomotor functions at a reasonable level of efficiency'. These considerations may affect the proportion of a fecund fish's body weight that is eggs and seem to predict an exponent of fecundity on W of 1.0. However, the argument of Hubbs, Stevenson & Peden (1968) is similar to that of this chapter. They found that in each of two fish species fecundity increased geometrically with body length. The exponent relating fecundity to body length was 'slightly less than two' (no data given). Hubbs *et al.* continued 'Although absorption would increase as does the square, muscle mass and other metabolizing tissues would increase with the cube of length increase. Therefore, maintenance needs would use larger and larger fractions of the absorbed nutrients as the fish grew longer. This differential could cause the fecundity changes to have an exponent of slightly less than two over most of the size range as well as preventing increase or even causing a decrease in the largest females.'

Conclusions

Intraspecifically, the energy females devote to reproduction increases with increasing body weight (invertebrate review by Spight & Emlen (1976)). This statement seems trivial, but it is argued in Chapter 7 that in species where males weigh more than females, increased male weight is associated with less energy available for reproduction.

A model is produced to predict the dependence of parental investment (measured in joules per unit time) on female body weight within species. It is shown that in species where females do not change much in weight over the course of their reproductive lifespans, the energy females devote to reproduction should scale on W with exponents of between 0.5 and 1.0. In species where each individual female reproduces on two or more occasions over a wide range of body weights, the energy invested in reproduction by females is predicted to scale on body weight more steeply than this. Data from 62 species of fish, five species of *Asellus* and two species of aphids are in agreement with this prediction. Of course, data from a variety of other taxa will need to be obtained before the predictions outlined here can be thought to hold generally.

5

Growth and productivity

In this chapter the allometric arguments introduced in Chapters 3 and 4 will be extended to growth curves, rates of reproduction and related topics.

It will be argued on theoretical grounds that

$$a > b$$

and evidence for this assertion will be reviewed. When values of a or b are required, they are taken as $a = 0.75, \frac{3}{4}, b = 0.67, \frac{2}{3}$, from Chapter 2. When the results depend critically on the values of a or b, the effects of $a \neq 0.75$, $b \neq 0.67$ are considered. Usually, however, the crucial requirements are that both a and b are less than 1, and that a exceeds b.

Growth curves

The three growth curves most frequently used for animals are the von Bertalanffy equation

$$\mathrm{d}W/\mathrm{d}t = AW^m - BW \tag{5.1}$$

the Gompertz equation

$$\mathrm{d}W/\mathrm{d}t = AW(\log_e W) \tag{5.2}$$

and the Logistic equation

$$\mathrm{d}W/\mathrm{d}t = AW - BW^2 \tag{5.3}$$

where A, B and m are constants, t is time and W is body weight (e.g. Schoener & Schoener, 1978; Alberch, Gould, Oster & Wake, 1979).

Just one of these, Equation (5.1), has a physiological basis. Von Bertalanffy (1960) assumed that catabolism – the breakdown of tissues – proceeded at a rate proportional to body weight, whereas the rate of anabolism – synthesis of tissues – was assumed to be proportional to surface area ($m = \frac{2}{3}$), body weight ($m = 1$), or some intermediate ($\frac{2}{3} < m < 1$). From Chapter 2, however, catabolism is not directly proportional to body weight. Consequently we might expect

$$dW/dt \propto k_2 W^b - k_1 W^a \tag{5.4}$$

(in the notation of Chapter 2) to give a better fit to data. In one sense, this would not be surprising: Equations (5.1) and (5.3) are particular forms of Equation (5.4), and Equation (5.4) includes more fitted constants than any of Equations (5.1), (5.2) or (5.3).

Unfortunately it does not appear as though Equation (5.4) can be solved analytically for all values of a and b (Hozumi, 1985). From Equation (5.4) we have

$$t \propto \int \frac{dW}{W^b - cW^a}$$

where $c = k_1/k_2$. Let $W^{a-b} = x$, then

$$t \propto \int \frac{1}{1 - cx} \cdot x^{(1-a)/(a-b)} \cdot dx$$

Let $cx = y$, then

$$t \propto \int \left(\frac{1}{1 - y} - \left(\frac{1 - y^{(1-a)/(a-b)}}{1 - y} \right) \right) \cdot dy$$

This can be solved if $(1 - a)/(a - b)$ is a natural number, because then

$$t \propto \int \left(\frac{1}{1 - y} - \frac{(1 - y)(1 + y + y^2 + \ldots + y^{(1-2a+b)/(a-b)})}{1 - y} \right) dy$$

and so

$$t \propto (-\log_e(1 - y) - y - y^2/2 - y^3/3 - \ldots - (a - b)/(1 - a)\, y^{(1-a)/(a-b)} + \text{constant}) \tag{5.5}$$

For $a = 1$, $b = \frac{2}{3}$, Equation (5.5) reduces to the most frequently cited form of the von Bertalanffy equation:

$$W = c_1(1 - c_2 e^{-c_3 t})^3$$

where c_1, c_2 and c_3 are constants.

For $a = \frac{3}{4}$, $b = \frac{2}{3}$, Equation (5.5) reduces to:

$$t = c_4(-\log_e c_5(1 - (k_1/k_2)W^{\frac{1}{12}}) - (k_1/k_2)W^{\frac{1}{12}} - (k_1/k_2)^2 W^{\frac{1}{6}} - (k_1/k_2)^3 W^{\frac{1}{4}})$$

where c_4 and c_5 are constants.

Ursin (1967, 1979) has estimated the values of the exponents a and b in Equation (5.4) for 81 fish species from a detailed review of their growth curves. He concludes that $a = 0.83$, $b = 0.59$ (with 95% confidence limits respectively equal to ± 0.06 and ± 0.02) give a significantly better fit than $a = 1$, $b = \frac{2}{3}$ as required by the most frequently used form of the von Bertalanffy growth equation. Equally, these values of a and b obviously give a better fit than $a = 2$, $b = 1$ as required by the Logistic growth equation, Equation (5.3).

In fish, intraspecific plots of log relative growth rate (percentage weight gained per day) on log body weight give straight lines with slopes that vary from -0.49 to -0.28 (Brett & Shelbourn, 1975; Brett, 1979). This means that as fish grow larger, each day their weight, although still increasing, increases by a smaller percentage of their current body weight. This can be predicted from Equation (5.4). As $a \simeq b \simeq 0.7$, we can write as an approximation

$$\mathrm{d}W/\mathrm{d}t \propto W^{0.7}$$

Therefore

$$(1/W)(\mathrm{d}W/\mathrm{d}t) \propto W^{-0.3}$$

Consequently a graph of log relative growth rate on log body weight is expected to have a slope of about -0.3.

Turning now from the intraspecific size dependence of relative growth rates to the interspecific scaling of growth rates (measured, for example, in grams per day), we can consider first the situation where most of the growth of an individual is caused by parental investment. This will be the case in many mammals, supplied as young mammals are by milk, and in some birds, for example many raptors and insectivorous birds. Here we expect maximal growth rates, measured as daily weight gains, interspecifically to scale at about $W^{0.7}$. This is because parental investment is predicted to scale at about $W^{0.7}$ per unit time (Chapter 2) on the assumption again that $a \simeq b \simeq 0.7$. (This is a simplication because in birds (Rahn *et al.*, 1975) and mammals (Tuomi, 1980) larger species often produce smaller litters. Consequently the interspecific slope of log growth rate on log body weight should slightly exceed 0.7.) Growth rates scale interspecifically as $W^{0.66}$ in birds, with 95% confidence limits of the exponent equal to ± 0.05 (Ricklefs, 1979), and as $W^{0.72}$ in mammals, where $r = 0.96$ (Case, 1978a).

For species where most growth occurs after the end of parental investment, maximal growth rates are again predicted to scale interspecifically at about $W^{0.7}$. From Equation (5.4), maximal growth rate occurs when $\mathrm{d}^2 W/\mathrm{d}t^2 = 0$. This occurs when

$$W^{a-b} = k_2 b / k_1 a \tag{5.6}$$

It will be argued in Chapter 7 that Equation (5.6) may represent the equation for optimal adult female body weight for a particular set of values of a, b, k_1 and k_2. Consequently for species where most growth occurs after the end of parental investment, maximal growth rates are predicted to occur as adulthood is approached. Maximal growth rates are therefore predicted interspecifically to scale in such species as $(k_2 W^b - k_1 W^a)$, i.e. at about $W^{0.7}$. Maximal growth rates scale interspecifically as $W^{0.61}$ in fish, where $r = 0.79$, and as $W^{0.67}$ in reptiles, where $r = 0.93$ (Case, 1978a).

Discussion

Whether Equation (5.4):

$$\mathrm{d}W/\mathrm{d}t \propto k_2 W^b - k_1 W^a$$

is more useful than other growth equations depends on its purpose. Physiologically it is probably more accurate than alternative existing equations, as shown above. It has, however, the disadvantage that it appears either difficult or impossible to solve analytically for all values of a and b. For some purposes, mathematical tractability is worth more than physiological relevance – plant growth equations, for example, are usually physiologically irrelevant (Hunt, 1978) – although this is perhaps less the case nowadays, with rapid and cheap numerical approximations available, than was once the case. An advantage of a physiologically relevant growth equation like Equation (5.4) is that it may suffice for a very large number of species. Additionally, extrapolations are less likely to lead to nonsensical results, which sometimes result from physiologically irrelevant equations.

The interspecific dependence of growth rates, dW/dt, and the intraspecific dependence of relative growth rates, $(1/W)(dW/dt)$, on size seem previously to have lacked explanations. In particular Ricklefs has repeatedly emphasized (Ricklefs, 1968, 1973, 1974, 1979) that 'the general decrease in growth rate, expressed as a percentage of adult weight, with increasing adult body weight still defies explanation' (quote from Ricklefs, 1974). Similarly, Case (1978a) writes 'Why, in fact, should growth rate and metabolic rate vary with body size at roughly the same rate? The answer is not at all obvious.'

Age at maturity, generation time and the duration of parental investment

When most of an individual's growth is caused by parental investment, growth rates, as shown above, are predicted to scale interspecifically at about $W_a^{0.7}$, where W_a is adult body weight. Consequently age at maturity (in units of time) will approximately be proportional to adult weight (in units of mass) divided by juvenile growth rate (in units of mass divided by time), that is

$$T_a \propto W_a / W_a^{0.7}$$

where T_a is age at maturity. We therefore have

$$T_a \propto W_a^{0.3} \qquad (5.7)$$

on the assumption that litter size is interspecifically only weakly dependent on body weight.

From Equation (5.7) it can also be predicted that generation time should scale interspecifically as $W^{0.3}$ in species where most of an individual's growth is caused by parental investment on the assumption that the gap in time between an organism reaching its adult weight and reproducing scales on body weight with an exponent of about 0.3, as appears to be the case (Taylor, 1965, 1968).

For species where most growth occurs after the termination of parental investment, Equation (5.4) holds and intraspecifically growth rate, $\mathrm{d}W/\mathrm{d}t$, is given by

$$\mathrm{d}W/\mathrm{d}t \propto k_2 W^b - k_1 W^a$$

As when we considered the scaling of growth rates and relative growth rates, we may approximate $a \simeq b \simeq 0.7$, so that

$$\mathrm{d}W/\mathrm{d}t \propto W^{0.7}$$

and therefore

$$\int_{W_o}^{W_a} \frac{\mathrm{d}W}{W^{0.7}} \propto \int_{T_o}^{T_a} \mathrm{d}t \qquad (5.8)$$

where W_o = weight at end of parental investment and T_o = age at end of parental investment. So

$$(T_a - T_o) \propto (W_a^{0.3} - W_o^{0.3})$$

In such species – where most growth occurs after the end of parental investment – we will have $W_a \gg W_o$. In Lepidoptera, for example, W_a/W_o varies from 500 to 10 000 (Moran & Hamilton, 1980). Consequently age at maturity, and similarly generation time, are again predicted to vary as approximately $W_a^{0.3}$.

From viruses to giant sequoia trees, generation time and age at maturity scale interspecifically from $W^{0.21}$ to $W^{0.33}$ (Bonner, 1965; Taylor, 1968; Fenchel, 1974; Finlay, 1977; Blueweiss *et al.*, 1978; Western, 1979; Baldock, Baker & Sleigh, 1980; Taylor & Shuter, 1981).

Predictions can now be made about the interspecific scaling of gestation time in mammals and incubation time in birds. In birds, egg weight scales interspecifically as $W^{0.68}$ (Rahn *et al.*, 1975). In mammals, neonate weight scales interspecifically as $W^{0.63}$–$W^{0.83}$ (Leitch *et al.*, 1959; Leutenegger, 1973, 1976). So in birds and mammals, offspring weight scales at about $W_a^{0.7}$. Because, as we would expect, prenatal growth rates scale, within developing embryos, at about $W^{0.7}$, at least in mammals (Huggett & Widdas, 1951; Payne & Wheeler, 1967; Frazer & Huggett, 1973, 1974), we have

$$\int_0^{W_a^{0.7}} \frac{dW}{W^{0.7}} \propto \int_0^T dt$$

where T is gestation time (for mammals) or incubation time (for birds). So

$$T \propto W_a^{0.21} \tag{5.9}$$

In mammals, gestation length scales interspecifically as $W^{0.24}$ with 95% confidence limits of the exponent equal to ± 0.02 (Millar, 1981). In birds incubation time scales as $W^{0.17}$ (Rahn *et al.*, 1975).

The interspecific ratio of litter mass to gestation time in mammals has been called 'reproductive growth rate' (e.g. Peters, 1983; Sibly & Calow, 1986). Given that, as above, litter mass in mammals scales interspecifically at about $W^{0.7}$, whereas Equation (5.9) predicts that gestation length should scale interspecifically at about $W^{0.21}$, we can expect the ratio of litter mass to gestation length interspecifically to scale at about $W^{0.7}/W^{0.21}$, that is at about $W^{0.49}$. In mammals the ratio of litter mass to gestation length interspecifically scales at about $W^{0.56}$–$W^{0.60}$ (Payne & Wheeler, 1968).

Finally, given data on fledgling weights in birds and weaning weights in mammals, we can predict age at fledgling in birds and duration of lactation in mammals. In a sample of 100 species of mammals Millar (1977) found that offspring weaning weight scaled interspecifically as $W^{0.73}$, with 95% confidence limits of the exponent equal to ± 0.04, $r = 0.95$. Similarly, Russell (1982) found that for 30 marsupial species litter weaning weight scaled interspecifically at $W^{0.71}$, $r = 0.81$. Rudder (1979), however, found that weaning weight in 16 haplorhine primate species scaled as $W^{0.95}$ as determined by the principle axis of correlation (Sokal & Rohlf, 1969), with 95% confidence limits of the exponent equal to ± 0.08, $r = 0.99$. In 8 strepsirhine primates weaning weight scaled interspecifically as $W^{0.97}$, with 95% confidence limits of the exponent equal to ± 0.10, $r = 0.99$. On the presumption, therefore, that weaning weights in mammals, and, by analogy, fledgling weights in birds scale interspecifically somewhere between $W^{0.7}$ and $W^{1.0}$, we expect duration of lactation in mammals and age at fledgling in birds (measured from hatching), by parallel arguments to those leading to the derivation of Equations (3.8) and (3.9), to scale interspecifically between $W^{0.21}$ and $W^{0.3}$.

In mammals the duration of lactation scales at $W^{0.15}$ as computed by Blaxter (1971), and as $W^{0.05}$ as computed by Millar (1977) – although this unusually low exponent, accompanied by the very low correlation coefficient of 0.20, is at least partly caused by the fact that the dataset included only two mammals over 4 kg in mass, the racoon, *Procyon lotor*, (5.5 kg), and the black-tailed deer, *Odocoileus hemionus*, (57 kg), and the white-tailed deer was incorrectly assigned an age at weaning of 21 days (lactation actually lasts for about 120–200 days (Mueller & Sadleir, 1977; Sadleir, 1980)). Rudder (1979) calculated an exponent of 0.56 for 22 haplorhines, with 95% confidence limits equal to ± 0.10, $r = 0.93$, and an exponent of 0.33 for 11 strepsirhines, with 95% confidence limits equal to ± 0.13, $r = 0.85$. In 57 marsupial species pouch life scales as $W^{0.28}$, $r = 0.89$ (Russell, 1982).

Discussion

There have been some attempts to explain how these variables depend on size. Fenchel (1974), writing r_m for reproductive rate, argued that 'The fraction r_m/(metabolic rate per unit weight) measures how much energy an organism spends for production relative to how much it spends for maintenance and this fraction can probably only vary within certain limits.' This explanation begs the question. First, why should the ratio of production to maintenance only vary within certain limits? (An explanation is suggested later in this chapter.) Secondly, r_m/(metabolic rate per unit weight) is not the same as the ratio of production to maintenance.

Southwood (1976) wrote that the positive correlation between size and generation time 'is probably due to longevity being inversely proportional to total metabolic activity per unit of body weight, and to the fact that the smaller the organism the greater the level of this activity.' Again, this explanation begs the question. To explain the positive correlation between size and generation time it is inadequate to state that larger species live longer. Furthermore, generation time depends far more on age at first reproduction than on longevity (Lewontin, 1965).

Many other cycle lengths besides those discussed in this section scale interspecifically at about $W^{0.25}$ (Lindstedt & Calder, 1981). Examples include the length of the cardiac cycle, the half-life of drugs (before they are broken down), and the time for a species to metabolize fat stores equal to 0.1% mass. Lindstedt and Calder use dimensional analysis (cf. Günther, 1975; Wilkie, 1977; Günther & Morgado, 1982) to argue that biological time should scale at $W^{0.25}$. The maximum stress (i.e. force divided by cross-sectional area) generated in homologous muscles is roughly constant (Hill, 1950). Therefore

$$\frac{\text{mass}\cdot\text{acceleration}}{\text{area}} \propto \frac{ld^2\cdot lt^{-2}}{d^2}$$

$$= l^2/t^2 = W^{\frac{1}{2}}/t^2 = \text{constant}$$

where l is length, d is diameter, W is weight and t is time. So t is proportional to $W^{0.25}$. Here Lindstedt and Calder have used McMahon's (1973) result (discussed in Chapter 2) that length is proportional to $W^{0.25}$. Lindstedt and Calder also use Prange's (1977) analysis to predict that t might scale between $W^{0.285}$ and $W^{0.33}$.

It is clear how such dimensional analyses predict that cycle lengths such as muscle contraction time should scale between $W^{0.25}$ and $W^{0.33}$, and Lindstedt and Calder argue that such analyses are sufficient to explain the dependence on W of all cycle lengths. It is difficult, however, to see how the dependence on W of variables such as generation time or duration of lactation are explained by dimensional analyses. Indeed the prediction of Equation (5.9) is that gestation time in mammals and incubation time in birds interspecifically should scale outside the range $W^{0.25-0.33}$, namely as $W^{0.21}$. More generally, if offspring weight at the end of parental investment scales interspecifically in a group of organisms as W^{θ} rather than specifically as $W^{0.7}$, the analysis leading to Equation

(5.9) would be replaced by

$$\int_0^{W_a^\theta} \frac{dW}{W^{0.7}} \propto T$$

that is

$$T \propto W_a^{0.3\theta}$$

The ratio of litter mass to gestation time would therefore be given by

$$W_a^\theta / T \propto W_a^{0.7\theta}$$

We can now see that the exponent relating 'reproductive growth rate' – that is, the ratio of litter mass to gestation length – interspecifically to adult body weight only equals 0.75 for values of θ close to 1.0. It has been generally expected that 'reproductive growth rate' should scale interspecifically on adult body weight with an exponent of close to 0.75 (Peters, 1983; Sibly & Calow, 1986). From the observation that it does not, Peters (1983) concluded that 'In other words, larger mammals devote a smaller proportion of ingestion, assimilation, or assimilation less respiration to reproduction than do small mammals', which is not in fact the case (Humphreys, 1979) whereas Sibly & Calow (1986) concluded that 'This suggests either that the costs of reproduction are more serious for larger animals and/or that they are less worth paying and this might be why iteroparity, rather than semelparity, is a more prominent feature of larger animals.'

Energy budgets and growth

As far as I know, direct measurements for the exponents a and b for the same species come from just two species. In the isopod *Idotea baltica*, a lies between 1.07 and 1.20, and b equals approximately 0.67 (Strong & Daborn, 1979). In the nematode *Plectus palustris*, a lies between 0.71 and 0.81 (Klekowski *et al.*, 1979), and b lies between 0.77 and 0.88 (Schiemer *et al.*, 1980).

Indirect information on the value of $a - b$ comes from the ontogenetic change in growth efficiency. Paloheimo & Dickie

(1966), Ursin (1967) and Staples & Nomura (1976) point out that the proportion of the energy budget that is devoted to growth will decrease ontogenetically if metabolic requirements scale more steeply on body weight than food intake does. Over a small unit of time, δt, we have from Equations (2.2), (4.1) and (5.4)

$$E_{gro} = (k_2 W^b - k_1 W^a)\delta t$$

where E_{gro} is the energy devoted to growth per unit time, and

$$E_{in} = k_2 W^b \delta t$$

so that we may write

$$\frac{E_{gro}}{E_{in}} = \frac{(k_2 W^b - k_1 W^a)\delta t}{k_2 W^b \delta t} \tag{5.10}$$

i.e.

$$\frac{E_{gro}}{E_{in}} = 1 - \frac{k_1}{k_2} W^{a-b} \tag{5.11}$$

where E_{gro}/E_{in} is the proportion of an individual's energy budget that is devoted to growth.

Growth efficiencies usually do decrease as an individual grows (Waldbauer, 1968; Calow, 1977; Table 5.1). Data for four species with extensive datasets are plotted in Figures 5.1(*a*)–(*d*). Analysis (Table 5.2) of the data in these figures shows that there is a closer correlation between $\log(1 - E_{gro}/E_{in})$ and $\log W$ than between $(1 - E_{gro}/E_{in})$ and W, as predicted by Equation (5.11).

From Equation (5.10) the proportion of the total energy budget that is devoted to growth from birth to adulthood, $\bar{E}_{gro}/\bar{E}_{in}$, can also be calculated. Consider an increase in weight from W to $W + \delta W$ over a period of time equal to δt. Then we have

$$\frac{\bar{E}_{gro}}{\bar{E}_{in}} = \frac{\int_0^W (k_2 W^b - k_1 W^a)\,\mathrm{d}W}{\int_0^W k_2 W^b \,\mathrm{d}W}$$

$$= 1 - \frac{k_1(1 + b)}{k_2(1 + a)} W^{a-b} \tag{5.12}$$

If we now make the major assumption, to be considered further in Chapters 6 and 7, that adult female body weight has evolved intraspecifically so as to maximize the energy that females can

Figure 5.1(*a*). Jersey cattle. Data from Brody (1945).

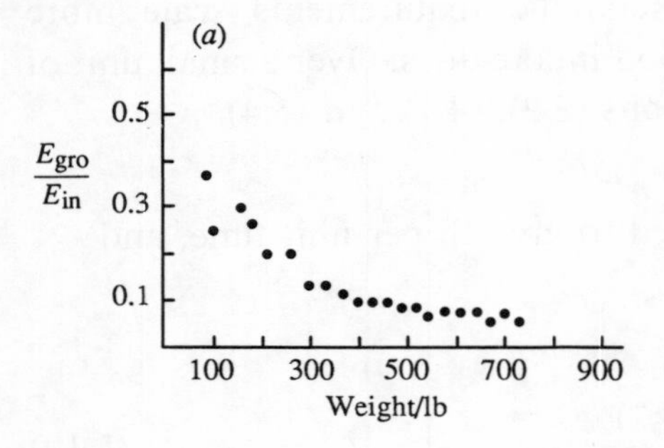

Figure 5.1(*b*). *Megalops cyprinoides*, fish. Data from Pandian (1967).

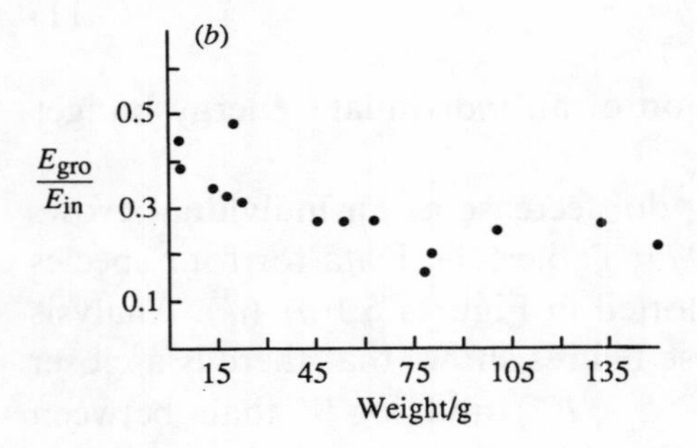

Figure 5.1(*c*). *Ophiocephalus striatus*, fish. Data from Pandian (1967).

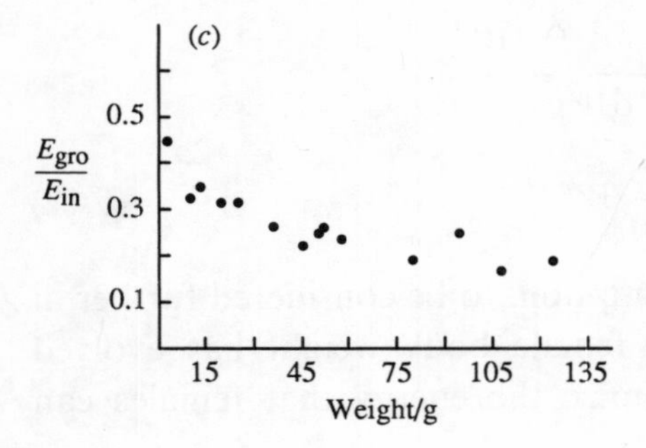

Figure 5.1(*d*). *Oceanodroma leucorhoa*, Leach's storm-petrel. Data from Ricklefs *et al.* (1980).

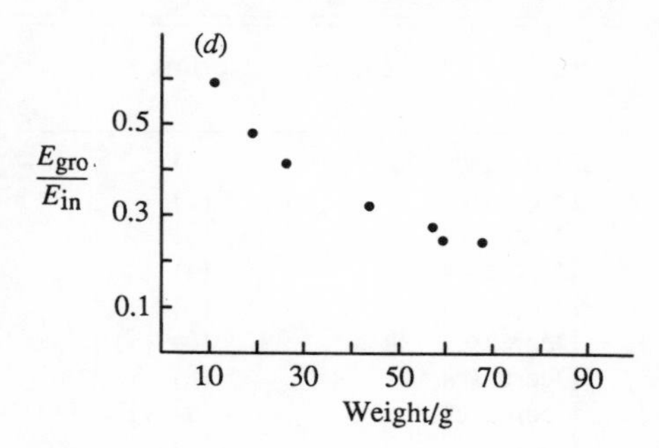

devote to reproduction, we have from Equation (3.2)

$$E_{rep} = k_2 W^b - k_1 W^a$$

and so

$$\frac{d}{dW}(k_2 W^b - k_1 W^a) = 0 \quad (5.13)$$

and

$$\frac{d^2}{dW^2}(k_2 W^b - k_1 W^a)|_{W=W_f} < 0 \quad (5.14)$$

where W_f is adult female body weight. Solving Equation (5.14) we have

$$W_f^{a-b} = k_2 b / k_1 a \quad (5.15)$$

(as in Equation (5.6)) and

$$a > b \quad (5.16)$$

We can now eliminate W_f from Equation (5.12) by substituting for W_f from Equation (5.15), whereupon we find

$$\frac{\bar{E}_{gro}}{\bar{E}_{in}} = 1 - \frac{b(1 + b)}{a(1 + a)} \quad (5.17)$$

Discussion

The direct measurements made of *a* and *b* do not provide convincing evidence that *a* exceeds *b*. The indirect evidence in Tables 5.1 and 5.2 suggest that *a* usually does exceed *b*. There are

Table 5.1. *The ontogenetic effect of body weight, W, on the proportion of its energy budget an individual devotes to growth, E_{gro}/E_{in}*

Species	Effect of increasing W on E_{gro}/E_{in}	Source
Bombyx mori, insect	Decreases	(1)
Agrotis orthogonia, insect	Decreases	(2)
Cimex lectularius, insect	Decreases	(3)
Phonoctonus nigrofasciatus, insect	Decreases	(4)
Lestes sponsa, insect	Decreases	(5)
Pleuronectus platessa, fish	Decreases	(6), (7)
Salmo trutta, fish	Decreases	(8)
Salvelinus fontinalis, fish	Decreases	(9), (10), (11)
Micropterus salmoides, fish	Decreases	(12)
Lepomis megalotis, fish	Decreases	(13)
Lepomis cyanellus, fish	Decreases	(13)
Limanda yokohamae, fish	Decreases	(14)
Kareius bicoloratus, fish	Decreases	(15)
Epinephelus guttatus, fish	Decreases	(16)
Cyprinodon macularius, fish	Decreases	(17)
Megalops cyprinoides, fish	Decreases	(18)
Ophiocephalus striatus, fish	Decreases	(18)
Histrio histrio, fish	Decreases	(19)
Pseudopleuronectus americanus, fish	Decreases	(20)
Ophiocephalus punctatus, fish	Decreases	(21)
Domestic chicken	Decreases	(22), (23), (24)
Calidris alpina, bird	Decreases	(25)
Sterna fuscata, bird	Decreases	(24)
Passer domesticus, bird	Decreases	(26)
Delichon urbica, bird	Decreases	(27)
Oceanodroma leucorhoa, bird	Decreases	(28)
Domestic cattle	Decreases	(22)
Laboratory rat	Decreases	(29)
Plectus palustris, nematode	No effect	(30)
Acyrthosiphon pisum, insect	No effect	(31)
Oncopeltus fasciatus, insect	No effect	(32)
Lygaeus palmii, insect	No effect	(32)
Salmo trutta, fish	No effect	(33)
Gadus morhua, fish	No effect	(34)
Salmo gairdneri, fish	No effect	(35)
Artemia salina, crustacean	Variable	(36)
Paropsis atomaria, insect	Variable	(37)
Dendrocygna autumnalis, bird	Variable	(38)
Stalia major, insect	Increases	(39)
Choristoneura fumiferana, insect	Increases	(40)
Tilapia mossambica, fish	Increases	(41)

Sources: (1) Hiratsuka, 1920; (2) McGinnis & Kasting, 1959; (3) Johnson, 1960; (4) Evans, 1962; (5) Fischer, 1972; (6) Dawes, 1930a; (7) Dawes, 1930b; (8) Pentelow, 1939; (9) Tunison, Phillips, McCay, Mitchell & Rodgers, 1939; (10) Phillips,

Sources for Table 5.1 (Cont.)

Tunison, Fenn, Mitchell & McCay, 1940; (11) Baldwin, 1956; (12) Prather, 1951; (13) Gerking, 1952; (14) Hatanaka, Kosaka & Satô, 1956a; (15) Hatanaka, Kosaka & Satô, 1956b; (16) Menzel, 1960; (17) Kinne, 1960; (18) Pandian, 1967; (19) Smith, 1973; (20) Chesney & Estevez, 1976; (21) Gerald, 1976; (22) Brody, 1945; (23) Medway & Kare, 1957; (24) Ricklefs, 1974; (25) Norton, 1970; (26) Blem, 1975; (27) Bryant & Gardiner, 1979; (28) Ricklefs, White & Cullen, 1980; (29) Mayer, 1948–1949; (30) Schiemer *et al.*, 1980; (31) Randolph, Randolph & Barlow, 1975; (32) Chaplin & Chaplin, 1981; (33) Brown, 1946; (34) Kohler, 1964; (35) Staples & Nomura, 1976; (36) Reeve, 1963; (37) Carne, 1966; (38) Cain, 1976; (39) Fewkes, 1960; (40) Koller & Leonard, 1981; (41) Rajamani & Job, 1976.

a variety of explanations possible for those occasions on which an ontogenetic decrease in growth efficiency has not been found. In some cases experimental error may have swamped any regular trend. This is particularly likely when the range of body weights observed was relatively small. In an early and detailed review Paloheimo & Dickie (1966) attributed the absence of any relationship between growth efficiency and size in brown trout, *Salmo trutta*, (Brown, 1946), and cod, *Gadus morhua*, (Kohler, 1964), to high scatter, and in neither case was there even a significant relationship between metabolic rate and body size. Pentelow's (1939) study on brown trout did reveal a significant decrease in growth efficiency with increasing size. Chaplin & Chaplin's (1981) study on the milkweed bugs *Lygaeus kalmii* and *Oncopeltus fasciatus* showed no ontogenic effect of W on E_{gro}/E_{in}. Their measurements, however, produced figures of 93.4 and 94.8% respectively for $\bar{E}_{gro}/\bar{E}_{in}$ from instars II to V inclusive. These figures are so high (Calow, 1977) that Chaplin and Chaplin's data must be called into doubt. In the fish *Tilapia mossambica* Rajamani & Job (1976) found an ontogenetic increase in growth efficiency, although they did suggest that this was most likely caused by the inability of the smallest fish to convert a type of food they are not at that stage normally used to.

If there really are species in which the proportion of the energy budget devoted to growth remains constant or even increases as the animal grows, then such species presumably show indeterminate growth. In the many species that do show determinate growth, the proportion of the energy budget that is devoted to growth must decline to zero in fully-sized adults.

Aside from Paloheimo & Dickie (1966), Ursin (1967) and Staples

& Nomura (1976), who pointed out that the proportion of the energy budget used for growth would decline with increasing intraspecific size if metabolic requirements scaled on body weight more steeply than food intake did, there have been many incorrect explanations for this phenomenon. Brody (1945), Prather (1951) and Gerald (1976) maintained that larger individuals have lower growth efficiencies because they have greater maintenance costs. This explanation fails to take into account the effect of size on food intake. Calow (1977) states that the phenomenon is associated with a progressive transformation of mitotic to functional tissue. Such a transformation presumably does take place, but by itself this explanation does not account for the precise allometric dependence of growth efficiency on size, which is supported by the data in Table 5.2.

Because growth efficiency decreases as individuals grow, farmers should slaughter their animals young. Under intensive farming methods this indeed happens.

Energy budgets and reproduction, and productivity/biomass ratios

Let us suppose again that female body weight has evolved so as intraspecifically to maximize the energy females can devote to reproduction, E_{rep}. Substituting the value of W_f given by Equation (5.15) into the equation for E_{rep} as a function of W_f, Equation (3.2), we have that

$$E^*_{rep} = k_1 \left(\frac{a}{b} - 1\right) W_f^a = k_2 \left(1 - \frac{b}{a}\right) W_f^b$$

where E^*_{rep} is the greatest intraspecific value of E_{rep}. By substituting for W_f into the equation relating energy assimilation to body weight, Equation (2.2), we also have

$$E_{in} = k_2 W_f^b = k_1 \frac{a}{b} W_f^a$$

and therefore

$$E^*_{rep}/E_{in} = 1 - \frac{b}{a} \tag{5.18}$$

Table 5.2. *Analysis of Figures 5.1a–d*

Species	Regression of $\log(1 - E_{gro}/E_{in})$ on $\log W$			Regression of $(1 - E_{gro}/E_{in})$ on W	Source
	$a - b$	95% confidence limits of $a - b$	r	r	
Jersey cattle	0.171	±0.025	0.958	0.900	(1)
Megalops cyprinoides, fish	0.072	±0.036	0.779	0.712	(2)
Ophiocephalus striatus, fish	0.087	±0.016	0.959	0.844	(2)
Oceanodroma leucorhoa, bird	0.329	±0.042	0.997	0.995	(3)

Sources: (1) Brody, 1945; (2) Pandian, 1967; (3) Ricklefs *et al.*, 1980.

Equation (5.18) is similar in value to, but lower than, the value of $\bar{E}_{gro}/\bar{E}_{in}$ given by Equation (5.17). For $a = \frac{3}{4}$, $b = \frac{2}{3}$, $(1 - b/a) = 0.11$, while $(1 - (b(1 + b))/(a(1 + a)) = 0.15$.

In general

$$1 \leqslant \left(1 - \frac{b(1 + b)}{a(1 + a)}\right) \Big/ \left(1 - \frac{b}{a}\right) \leqslant 2 \qquad (5.19)$$

Neither Equation (5.17) nor Equation (5.18) includes W_f. Consequently both E^*_{rep}/E_{in} and $\bar{E}_{gro}/\bar{E}_{in}$ are predicted to be independent of species weight. E^*_{rep} and $\bar{E}_{gro}$ are often combined in reports of energy budgets as 'production'. Humphreys (1979), in an analysis of 235 energy budgets, shows that P/A, the ratio of production to assimilation, is indeed independent of species weight, where assimilation equals production plus respiration. The exponent of the least squares common slope of log production on log respiration equalled 0.96, with 95% confidence limits equal to ± 0.04, $r = 0.94$.

We would expect productivity per individual per unit time to scale interspecifically at about $W_a^{0.7}$, as these are the predictions both for reproduction (Chapter 3) and growth (above). Productivity/biomass ratios, P/B, are often calculated by ecologists (reviewed by Banse & Mosher, 1980). Productivity equals the number of individuals sampled, N, multiplied by each individual's productivity. Biomass equals N multiplied by the weight of each individual.

Consequently we predict

$$P/B \propto (W_a^{0.7} \cdot N)/(W_a \cdot N)$$

and therefore

$$P/B \propto W_a^{-0.3} \qquad (5.20)$$

In invertebrates P/B scales interspecifically as $W^{-0.37}$, with 95% confidence limits of the exponent equal to ± 0.07, $r = 0.91$ (Banse & Mosher, 1980). In fish P/B scales interspecifically as $W^{-0.26}$, with 95% confidence limits of the exponent equal to ± 0.16, $r = 0.77$ (Banse & Mosher, 1980). In mammals P/B scales interspecifically as $W^{-0.27}$, with 95% confidence limits of the exponent equal to ± 0.03, $r = 0.93$ (Farlow, 1976).

'Turnover ratio' is defined by Waters (1969) as the ratio of a cohort's production to the mean standing crop measured over an entire single life cycle. We would expect this ratio to scale interspecifically on W with an exponent close to zero on the grounds that productivity should scale on W with an exponent of about 0.7, whereas life cycle scales on W, as considered earlier, with an exponent of about 0.25. Waters (1969) reviewed data on turnover ratios and concluded that '. . . they exhibit a remarkable degree of constancy, varying only between about 2.5 and 5'.

Discussion

There do not seem to have been any previous explanations suggested for why P/A, the ratio of production to assimilation, is interspecifically independent of body weight, nor do there appear to have been any previous explanations suggested for why P/B scales interspecifically at about $W^{-0.3}$ with the exception of Peters (1983), who pointed out that we would expect P/B to scale at about $W^{-0.25}$ on the grounds that the unit of P/B is the reciprocal of time. Koller & Leonard (1981) even thought that P/A ratios should be smaller in smaller species on the grounds that metabolic rate and body size are inversely related. As far as I know there have been no previous predictions for the scaling of 'turnover ratio' (*sensu* Waters, 1969).

An important point arising from the fact that Equations (5.17) and (5.18) are independent of W_a is that interspecifically we expect energy requirements and energy assimilation to scale on size with the same exponents even though intraspecifically we expect the slope of log energy requirements on log weight to be steeper than the slope of log energy assimilation on log weight.

At first sight, it might appear that Equations (5.17) and (5.18), which predict production to assimilation ratios of between 11% and 15% for $a = \frac{3}{4}$, $b = \frac{2}{3}$, predict Lindeman's 'Law of Trophic Efficiency', which states that in ecosystems the efficiency of energy transfer from one trophic level to the next is about 10%. In fact, as May (1979) has pointed out, to determine trophic efficiencies, two questions must be answered. First, what fraction of the net productivity at one trophic level is actually assimilated by creatures at the next level? Secondly, how do these organisms apportion their assimilated energy between net productivity and respiration? Equations (5.17) and (5.18) relate only to the second question. We lack any quantitative predictions of the fraction of the net productivity at one trophic level that is actually assimilated by the next level. In general, trophic efficiency equals the exploitation efficiency (the percentage of production by one trophic level consumed by the next) multiplied by the assimilation efficiency (the proportion of the consumed food that is assimilated) multiplied by the production efficiency (which equals the proportion of the assimilated energy that ends up in growth and reproduction). Actually, trophic efficiencies, rather than being fairly constant at around 10%, vary considerably. In passing, it may be worth noting that Lindeman (1942), although he was the first to calculate trophic efficiencies (which ranged from 5.5% to 22.3% for the two ecosystems he studied), was too cautious to suggest his results could be extrapolated to other ecosystems. He himself never suggested any law of Trophic Efficiency.

A major shortcoming of this section is that it fails to account for the very low values of P/A found in homeotherms, or for the very high values of P/A found in some poikilotherms. P/A ratios vary from 0.86%, on average, in mammalian insectivores, to 55.6%, on average, in carnivorous non-social insects (Humphreys, 1979). Why it is that homeotherms have lower P/A ratios than poikilotherms is an intriguing question. It seems almost universally to have been thought a straightforward consequence of the greater metabolic rates of homeotherms (McNeill & Lawton, 1970; Turner, 1970; May, 1979; Lawton, 1981; Lavigne, 1982). However, this explanation fails to take into account the fact that homeotherms also have much greater intake rates than similar-sized poikilotherms (data in Farlow, 1976).

The extents to which homeotherms have greater rates of energy assimilation and energy expenditure are logically connected to the relative production efficiencies of the two groups. Figure 5.2 shows Hemmingsen's famous comparison of metabolic rates in homeotherms, poikilotherms and unicellular organisms. Hemmingsen produced three equations relating basal metabolic rate, BMR, to body weight, W:

$$BMR_{\text{unicellular organisms}} = 0.0176W^{0.756}$$

$$BMR_{\text{poikilotherms}} = 0.144W^{0.738}$$

$$BMR_{\text{homeotherms}} = 4.10W^{0.739}$$

where BMR was measured in cubic centimetres of O_2 per hour and W in grams. As $0.738 \simeq 0.739$, we have that $4.10/0.144 = 28.5$ gives the ratio of the rate at which a homeotherm metabolizes relative to a poikilotherm of the same mass. This disparity would probably be found to be even greater were adequate data available for the scaling of average daily metabolic rate both in homeotherms and poikilotherms (Chapter 2). Now from Humphreys (1979), we have that homeotherms have productivity/assimilation ratios of about 2.5%, whereas poikilotherms have productivity/assimilation ratios of around 25%.

For homeotherms we can therefore write

Figure 5.2. Standard metabolic rates of homeotherms, poikilotherms and unicells. Taken from Peters (1983).

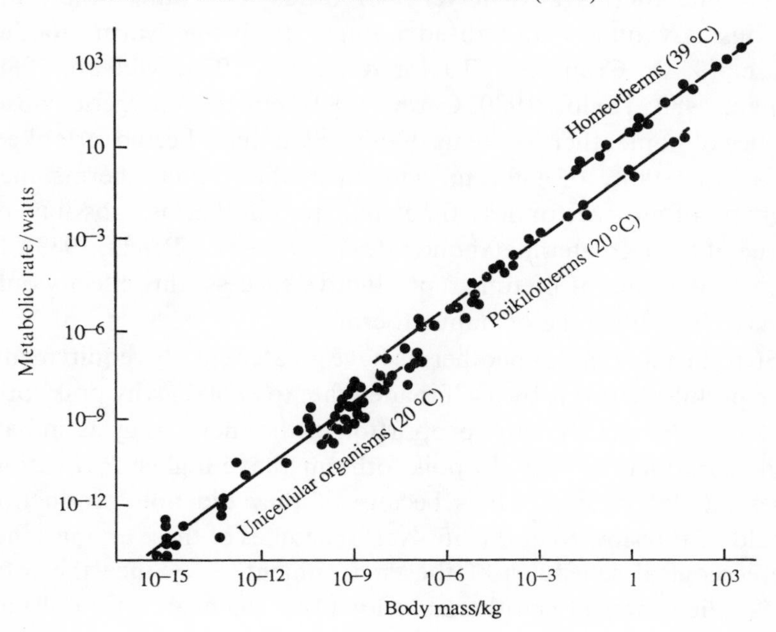

$$(E_{in} - E_{req})/E_{in} \simeq 0.025$$

whereas for poikilotherms

$$(E_{in} - E_{req})/E_{in} \simeq 0.25$$

Substituting E_{req} for a homeotherm equal to $28.5E_{req}$ for a poikilotherm, we find that homeotherms assimilate energy approximately 22 times as quickly as poikilotherms. The fact that homeotherms have lower P/A ratios than poikilotherms therefore rests on the fact that 22 is less than 28.5. Theoretically it is quite possible for a homeotherm to have a greater P/A ratio than a poikilotherm, provided that the extra rate at which it assimilates energy more than compensates for the extra rate at which it expends energy on everything except growth and reproduction.

It is worthwhile remarking that natural selection does not, of course, work on production efficiencies. The difference between energy assimilation and non-reproductive energy requirements is far more likely to be of importance as this energy can be channelled into reproduction. This difference is substantially greater for homeotherms than for poikilotherms of the same size, despite the fact that homeotherms have lower P/A ratios. Of course, there are various advantages and disadvantages to being warm-blooded (Case, 1978b; Crompton, Taylor & Jagger, 1978; McNab, 1980; Pough, 1980; Taylor, 1980; Carey, 1982), but this energetic consequence of homeothermy seems previously to have been overlooked. It is important, though, to remember that homeotherms may perhaps often live for less time than poikilotherms, possibly by virtue of their greater metabolic rates (reviewed by Peters, 1983). In terms, therefore, of lifetime reproductive success, this effect would reduce the advantage of homeothermy.

So if the fact that homeotherms have greater energy requirements than poikilotherms is by itself inadequate to explain why poikilotherms can channel a greater proportion of the energy they assimilate into reproduction, why do poikilotherms have higher P/A ratios? Wieser (1985) suggests it is because if they did not 'production would be too slow to fit the life cycle schedules of these animals into the ecological schedules of the environment. . . . Since the total metabolic power of ectotherms is low, the synchronization of biological and ecological cycles seems to be possible only by accelerating the reproductive processes and by channeling a large share of the metabolic power to the site of reproduction.' The trouble with this theory is that it is difficult to see the disadvantage to a homeotherm similarly of channelling a large share of the food assimilated to reproduction.

Various other explanations can tentatively be suggested for why poikilotherms are energetically more efficient than homeotherms. One possibility is that poikilotherms may spend a greater proportion of their life growing. As considered above, growth efficiencies are energetically greater than reproductive efficiencies. A second possibility is that some poikilotherms may start to reproduce well before they reach their full adult size. In this case Equation (5.18) would underestimate the proportion of the energy assimilated that can be devoted to reproduction because, as noted above (Equation (5.11)), the proportion of the energy budget that can be devoted to growth and reproduction declines as an individual gets larger. A third possibility is that outside the breeding season and when food may be in short supply, poikilotherms probably have only to metabolize a little of their stored reserves to stay alive. For warm-blooded organisms, however, winter may be a time when a great deal of energy has to be expended simply to ensure survival – hibernation is a strategy to reduce the magnitude of this effect. Consequently, P/A ratios, when measured over an organism's lifespan, would be expected to be lower for mammals and birds that live for at least one year than for both poikilotherms, whatever their lifespans, and the few homeotherms that survive for less than a year. A fourth possibility, although it seems less likely than the above three, is that poikilotherms have lower values of b/a than homeotherms. Were this to be the case, we would expect, from Equation (5.18), homeotherms to have lower values of E_{rep}/E_{in} than poikilotherms.

This section may shed some light on the great range in weights of animals that humans farm. At one extreme of husbandry is intensive meat or egg production. Here farmers provide livestock with all their food requirements: profitable farmers wish, approximately, to maximize P/A. P/A is predicted to be independent of species weight.

At the other extreme are farmers who provide their animals with only a small proportion of their food requirements, as, for example, is often the case with grouse and deer. Such farmers wish, approximately, to maximize food output per unit area, i.e. productivity multiplied by population density. Productivity is proportional interspecifically to about $W^{0.7}$ (Chapter 3), whereas population density, at least in herbivorous mammals (Damuth, 1981a) scales interspecifically as $W^{-0.7}$, with 95% confidence limits of the exponent equal to ± 0.08, $r = 0.86$. Consequently food output per unit area per unit time is, like P/A, predicted to be independent of species weight. Of course larger species may be overexploited more easily than smaller ones, and take longer to recover. Consequently we would predict given overfishing, for example, that larger species should cost more per kilogram than smaller ones. This is the case for fish in the Great Lakes of the USA (Regier, 1973).

Why are there so many species?

From observations of the physical dimensions of closely related species of animals living in the same habitat, Hutchinson (1959) suggested that for two species to coexist in different niches but at the same level of a food web, characters related to the trophic apparatus need to differ in length, in the two species, by a ratio of approximately 1.3. Subsequently May & MacArthur (1972) predicted that the average food sizes for species adjacent on a one-dimensional resource food continuum of food size must differ by an amount roughly equal to the standard deviation in the food size taken by either species. May & MacArthur's work has been extended to more than one dimension, but with similar results, by Rappoldt & Hogeweg (1980). Maiorana (1978) has predicted Hutchinson's findings by combining May & MacArthur's (1972) theory with the observed coefficients of variation for animal structures.

The point of this section is to argue that species' rates of reproduction are so size-dependent that to ignore them, as May &

MacArthur and Maiorana do, is to render inadequate attempts quantitatively to explain species coexistence.

The number of offspring an individual of body weight W produces, per unit time, scales interspecifically as $W^{0.7}/f(W)$, where $f(W)$ is the energy invested in each offspring, because E_{rep} scales interspecifically at about $W^{0.7}$ (Chapter 3). Because E_{req} and E_{rep} scale interspecifically at about $W^{0.7}$ (Chapters 2 and 3), we would predict that population density would scale interspecifically at about $W^{-0.7}$ on the assumption, considered below, that the amount of food available to a species per unit area is independent of W. In herbivorous mammals population density scales at $W^{-0.70}$ (Damuth, 1981a). Therefore the number of offspring produced per unit area per unit time by a species of body weight W scales at $1/f(W)$. Usually $f(W)$ increases with increasing W. Neonate weight scales as $W^{0.68}$ in birds, with 95% confidence limits of the exponent equal to ± 0.02 (Rahn *et al.*, 1975); as $W^{0.63}$–$W^{0.83}$ in mammals (Leitch *et al.*, 1959; Leutenegger, 1973, 1976); as $W^{0.42}$ in reptiles, $r = 0.84$ (Blueweiss *et al.*, 1978); as $W^{0.43}$ in fish, $r = 0.51$ (Blueweiss *et al.*, 1978); and as $W^{0.24}$ in Crustacea, $r = 0.59$ (Blueweiss *et al.*, 1978). In 23 species of Ecudorian frogs, however, egg weight was found to be independent of adult female body weight (Crump, 1974). Were species survival determined by the number of individuals produced per unit time per unit area, smaller species would typically enjoy a great advantage.

Of course larger species enjoy many advantages (Clutton-Brock & Harvey, 1983), but it does not yet seem possible to generalize these quantitatively. Simple allometric arguments are unlikely to tell us very much about species coexistence. This view contradicts Damuth (1981a). Damuth argues that 'The energy used by the local population of a species equals the population density (D) multiplied by individual metabolic requirements (R), which yields the following relationship to body mass (W): $DR \propto W^{-0.75} W^{0.75}$ . . . The independence of species energy control and body size revealed by this reciprocal relationship implies that random environmental fluctuations and interspecific competition act over evolutionary time to keep energy control of all species within similar bounds.'

Discussion

Gause's competitive exclusion principle states that two species survive indefinitely in coexistence only when they occupy different niches in the microcosm in which they have an advantage over their competitors (cf. Gause, 1934). Some regard the competitive exclusion principle as the most important development in theoretical ecology; others dismiss it as a trite maxim (reviewed by Krebs, 1972). May & MacArthur's (1972) paper was so important because it predicted what degree of overlap would still allow two species to coexist. May & MacArthur carefully considered the robustness of their results (although see Heck, 1976), but their fundamental premise is that the required niche separation between two species is necessary to allow coexistence in an environment where fluctuations in population number are entirely stochastic. In contrast, size-dependent rates of reproduction are deterministic. This criticism is particularly important when May & MacArthur's model is used, as by Maiorana (1978), to predict size differences in species occupying contiguous niches, and segregated only by virtue of their size difference. Hutchinson's (1959) observations still lack an explanation, although the evidence in support of his generalization has been strongly criticized (Roth, 1981; Simberloff & Boecklen, 1981).

Home range sizes and *r* and *K* selection

The first quantitative study of home range sizes was provided by McNab (1963a). McNab collated data on the home range sizes of 26 species of mammals. These 26 species were divided into two groups – 'hunters' and 'croppers'. Hunters were species that were granivorous, frugivorous, insectivorous or carnivorous. Croppers were species that grazed or browsed. McNab then produced fitted equations for home-range size, HRS, measured in acres, as a function of weight, W, in kilograms, separately for hunters and croppers:

$$\mathrm{HRS}_{\mathrm{croppers}} = 3.02W^{0.69}$$

$$\mathrm{HRS}_{\mathrm{hunters}} = 12.6W^{0.71}$$

McNab pointed out that the exponents relating home-range size to body weight, 0.69 and 0.71, did not differ significantly from the exponent of 0.75 shown by Kleiber (1961) to relate basal metabolic rate to body weight. McNab noted though that what is important for animals in nature is not their basal rate of energy expenditure, but, rather, the total daily expenditure. In the absence, at that time, of almost any data on daily energy expenditure in nature, McNab concluded that 'The size of the home range in mammals, accordingly, is determined by the rate of metabolism'. Larger mammals have larger home ranges because they need more energy. Croppers have smaller home ranges than hunters because of the greater concentration of their food within a given area.

Subsequent to McNab's work, Schoener (1968) found that the exponents relating territory size to body weight in birds differed significantly between predators and herbivores. For birds that fed on 90–100% animal food, the exponent was 1.31. For birds that fed on 0–10% animal food, the exponent was only 0.70. To see whether the difference between his results and McNab's suggested some difference between birds and mammals, Schoener analysed data collected since the publication of McNab's paper and found that the exponent relating home-range area to the body weights of nine species of non-subterranean terrestrial predatory mammals was 1.41, almost the same as the exponent of 1.39 calculated by Schoener for the exponent relating home-range area interspecifically to body weight in predatory birds. Schoener concluded that territory or home-range size increases more rapidly with body weight for predators than for herbivores in both birds and mammals. He suggested that this reflected a rapidly decreasing food density for predators of increasing weight.

Recently, in a careful analysis of the interspecific scaling of home-range size in 23 species of mammals for which the same procedure could be used for all species to determine home-range size, Swihart, Slade & Bergstrom (1988) found that Schoener's division between species at different trophic levels no longer held. McNab's classification was used to divide the species into hunters and croppers, and the following regressions were computed:

$$\mathrm{HRS}_{\mathrm{croppers}} = 4.90W^{1.56}$$
$$\mathrm{HRS}_{\mathrm{hunters}} = 15.14W^{1.26}$$

where home-range size, HRS, is in hectares and body weight, W, is in kilograms. The exponents of 1.56 and 1.26 do not differ significantly.

The results of Swihart *et al.* (1988) suggest that the exponents relating home-range area to body size may not differ between herbivores and carnivores. They also suggest that the exponents are greater than the exponents relating basal metabolic rate to body size. Why might this be? We can rule out the possibility that average daily metabolic rate might scale on body weight more steeply than does basal metabolic rate. The exponents are very similar (Chapter 2) and significantly less than one.

A second possibility, as originally suggested by Schoener for predators, is that the productivity of the environment might scale negatively with size, so that larger species require larger feeding areas relative to their metabolic requirements. Harestad & Bunnell (1979) postulate that large mammals experience a greater proportion of energetically useless space within their home range because their resources are more patchily distributed. The major problem with this explanation is that it seems totally *a posteriori.* Had the exponent relating home-range area to body weight been less than 0.75, one could imagine authors suggesting that smaller animals, with their relatively high metabolic rates, require high-energy foods that are more patchily distributed than the foods of larger species.

A third hypothesis has been suggested by Damuth (1981b), who argues that in larger species the home-range area is shared with more conspecifics (Figure 5.3). Where S is the average number of individuals of a species that live in an area the size of the average individual home range of the species, Damuth determined S as a function of W, in grams, for 18 species of herbivorous mammals:

$$S = 0.44W^{0.34}$$

Even if habitat productivity was independent of species size, Damuth's result would predict that the exponent relating home-range size to body size would be greater than the exponent relating metabolic requirements to body size simply because in larger species the home-range area is shared by more conspecifics. I find it difficult, however, to imagine that Damuth's explanation could explain the steep slope of log feeding territory size on log body weight for predatory birds.

Finally, Lindstedt, Miller & Buskirk (1986) argue that we should not be concerned with *daily* metabolic requirements. To do so, they argue, is to confuse chronological time with biological time. What is important to an organism, they maintain, is not its daily energetic needs, but its energetic needs over some biologically relevant period such as gestation or lactation, time to reach independence or adult size, or even lifespan itself. Each of these measures scales at approximately $W^{0.25}$, so that the total energetic requirements over biological time periods scale at about $W^{1.0}$. It could be argued, however, that Lindstedt *et al.* (1986) seem not to have taken into account the importance of the regeneration of the food supply in a feeding area over time.

It is evident that we are still a considerable way from understanding why animals have the home ranges that they do (Reiss, 1988).

Figure 5.3. *S*, the average number of individuals of a species that live in an area the size of the average individual home range of the species, as a function of *W*, body mass in grams, for 18 species of herbivorous mammals. Taken from Damuth (1981b).

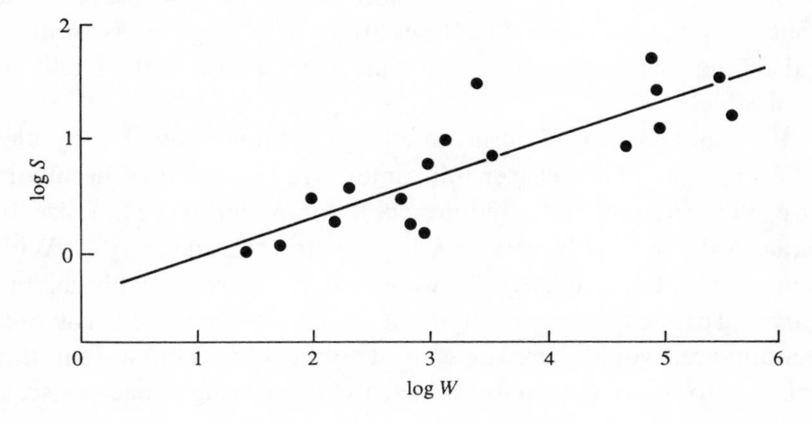

It is clear that considerable theoretical and empirical work needs to be done. Additionally, regression coefficients depend on the quality of the data used, the taxonomic level at which the analyses are performed and the regression techniques used (Chapter 1). I suspect that we presently know less about the scaling of home-range area than about the dependence on size of almost any other ecological, anatomical or behavioural variable.

Can allometry tell us anything about the scaling on size of r, the natural rate of increase, and K, carrying capacity?

An equation for r is:

$r \propto$ (generation time)$^{-1}$ $\times$ (number of offspring per individual per generation)

Now the number of offspring produced per individual per generation does not scale on body weight in different taxa in a consistent manner. In mammals and birds, larger species have smaller clutches, but live longer. The combined result is that the number of offspring produced per generation is perhaps almost independent of species size. Accordingly, r is predicted to scale on W as generation time scales on W, that is as about $W^{-0.25}$. In poikilotherms, however, larger species produce many more eggs per clutch, fecundity scaling at about $W^{0.47\text{–}0.70}$ (Blueweiss *et al.*, 1978), and possibly live for longer. Accordingly, one would expect larger species of poikilotherms to be more r-selected! This flies in the face of conventional wisdom; r has been found to scale on W with an exponent of -0.28 (Fenchel, 1974) or -0.26 (Blueweiss *et al.*, 1978). It is true that the data used in these analyses are very heterogeneous. Statistical re-analysis within taxa with only high quality data might be profitable.

If we assume that plant productivity (in units of available joules of energy per unit area per unit time) is independent of herbivore size, then the metabolic requirements for N herbivores of size W equal NW^a per unit area and K is predicted to scale at W^{-a}. With a approximately equal to 0.75, we would therefore expect herbivore carrying capacities to scale at about $W^{-0.75}$. The mass of herbivores per unit area would therefore scale at about $W^{0.25}$. At first sight, this appears to imply that first-level carnivore carrying capacities scale at about $W^{-0.5}$, as per unit area N carnivores would have a total metabolic requirement proportional to $NW^{0.75}$ and a volume of food that scales at $W^{0.25}$. This, however, is to ignore the regeneration of the carnivores' food supply. If herbivores replace themselves at a rate that scales at $W^{-0.25}$, then herbivore productivity (in units of available joules of energy per unit area per unit time) would be independent of herbivore size, as was the case for plant productivity, and carnivore carrying capacity, K, would also scale at $W^{-0.75}$. In a detailed review Peters & Wassenburg (in Peters, 1983) found that animal density scaled at between $W^{-1.17}$ and $W^{-0.52}$.

The cost of play

Animal play is virtually confined to birds and mammals, and within these two groups occurs mainly in juveniles. The benefits and costs of play are difficult to quantify, and many have been suggested (Fagen, 1981). Martin (1982) has suggested a definition of the energy cost of play and provides an estimate of its likely magnitude. The aims of this section are to translate such estimates into the selective coefficients favoured by population geneticists, and to suggest a new explanation for why play occurs predominantly in juveniles.

Martin defined the energy cost of play (ECP) as the net daily energy expenditure (in excess of resting metabolism) due to play, expressed as a percentage of the total daily energy budget. So

$$\mathrm{ECP} = t_{\mathrm{p}}(\mathrm{PMR} - \mathrm{RMR})/\mathrm{ADMR} \tag{5.21}$$

where ADMR is average daily metabolic rate, t_{p} is the percentage of total time spent playing, PMR is the mean metabolic rate when playing, and RMR is the resting metabolic rate.

Following Martin, this analysis is restricted to the energetic cost of play. As Martin and others have pointed out, play may also have survivorship costs resulting from increased costs of injury, predation, etc., but these cannot as yet be quantified. Again following Martin, it is assumed that the alternative to play is resting. This makes the implicit assumption that feeding time (strictly energy assimilation) would remain unaltered were play to cease.

For adults, where a reduction in play may result in more energy being available for reproduction, the reproductive cost of play can be estimated. In the absence of play, the proportion of the energy budget used by an adult for reproduction, REP, would be increased by an additive factor of ECP on the assumption that the energy used for play could be devoted instead to reproduction.

For juveniles, a reduction in play may result in more energy being available for growth. Consider a juvenile of weight W. During a time interval of δt, weight gain, δW, will be given by

$$\delta W \;=\; C(k_2 W^b \;-\; k_1 W^a)\delta t \tag{5.22}$$

where C is a constant that relates energy stored to weight gain. For a hypothetical playless individual, k_1 will be reduced by a factor x and Equation (5.22) becomes

$$\delta W \;=\; C(k_2 W^b \;-\; (k_1/x)W^a)\delta t \tag{5.23}$$

From Equation (5.21) we can relate ECP and x. We have

$$\mathrm{ECP} \;=\; (k_1 W^a \;-\; (k_1/x)W^a)/k_1 W^a \tag{5.24}$$

From Equations (5.22)–(5.24) we have that the ratio of the weight gained by a playless individual to the weight gained by a playing individual equals

$$1 + k_1 W^a \mathrm{ECP}/(k_2 W^b \;-\; k_1 W^a)$$

which can be written as

$$1 + \{(E_{\mathrm{in}}/E_{\mathrm{gro}}) \;-\; 1\}\mathrm{ECP} \tag{5.25}$$

Discussion

Relative to a 'wild-type' playing individual with fitness = 1, an adult 'mutant' playless individual has a fitness, $1 + s$, given by

$$1 + s \;=\; (\mathrm{REP} + \mathrm{ECP})/\mathrm{REP} \;=\; 1 + (\mathrm{ECP}/\mathrm{REP}) \tag{5.26}$$

on the assumption that fitness is proportional to the energy invested in reproduction. This assumption is defended in Chapter 7. In a survey of 70 mammalian energy budgets, Humphreys (1979) found that REP typically lay between 1 and 10%, with a mean of 2.76%. Using data from the literature Martin (1982) concluded that a realistic estimate of ECP for a juvenile mammal would be 2.5%. These figures lead to a value of s in Equation (5.26) equal to 0.91. Accordingly, it is perhaps unsurprising that adults play little. Equation (5.25) shows that in juveniles a given amount of play has a greater effect on weight gain as E_{gro}/E_{in} decreases, i.e. as individuals grow. We therefore expect play to be less common in older individuals.

An alternative explanation for why play is primarily a behaviour of juveniles is provided by Fagen (1977) who assumes that play has immediate costs, in terms of growth and survivourship, but delayed benefits, in terms of growth, survival and fecundity, all at later ages. In all his simulations the incidence of play decreased monotonically with increasing age. This conclusion follows automatically from his assumption that play has delayed benefits at all later ages. Under this assumption, play is obviously more advantageous the sooner it occurs. Fagen may be quite correct in this assumption; the above analysis suggests that play may become more costly, as well as less beneficial, in older animals. It is not the case that play begins literally at $W = 0$. Rather, as Fagen (1976) points out, play begins when the young animal acquires the strength and coordination to exhibit adult motor patterns.

These quantitative conclusions rely on the assumptions that the alternative to play is resting, and that energy intake would remain the same were play to cease. The degree to which these assumptions are violated is unknown, but is unlikely quantitatively to affect the conclusions, and if known could be incorporated into the calculations.

Conclusions

The allometric model introduced in Chapter 3 is extended to consider growth curves and new explanations are suggested for the following phenomena:

the intraspecific dependence of relative growth rate (percentage weight gained per day) on body weight;

the interspecific scaling of growth rate (absolute weight gained per unit time, e.g. grams per day) on species size;

the interspecific scaling on adult body weight of age at maturity, generation time, gestation time, incubation time, the ratio of litter mass to gestation time (sometimes called 'reproductive growth rate'), age at fledgling and duration of lactation;

the ontogenetic change in growth efficiency as individuals increase in weight;

the interspecific dependence of production/assimilation and production/biomass ratios on body weight;

the interspecific scaling on adult body weight of the ratio of a cohort's production to the mean standing crop measured over an entire single life cycle (sometimes called 'turnover ratio').

The reasons why poikilotherms have higher production/assimilation ratios than homeotherms is not a straightforward consequence of the metabolic cost of homeothermy. Homeotherms also assimilate energy much more quickly than do poikilotherms. Natural selection does not work on production efficiencies. The difference between energy assimilation and non-reproductive energy requirements is far more likely to be of importance as this energy can be channelled into growth and reproduction. This difference is substantially greater for homeotherms than for poikilotherms of the same size. Various explanations can tentatively be proposed for why poikilotherms are energetically more efficient than homeotherms.

The size-dependence of rates of reproduction renders inadequate current quantitative theories of species coexistence, although simple allometric arguments are unlikely to tell us very much about species coexistence.

We still do not understand why animals have the home-range areas that they do. In particular, the interspecific exponents relating home-range areas to body weight are considerably steeper in those taxa examined than are exponents relating energy requirements to body weight.

It has been believed that r, the natural rate of increase, scales interspecifically on body weight with an exponent of about -0.25, so that, as seems obvious, larger species are less r-selected. It is suggested here, however, that in poikilotherms, larger species may be *more* r-selected. It is predicted that carrying capacity, K, should scale at about $W^{-0.75}$, irrespective of the trophic level of the organisms considered.

Finally, the cost of play is considered, and a new reason suggested for why play is mainly an activity of juveniles.

6

Quantitative models of body size

In this chapter existing quantitative models of body size are critically reviewed. Special attention is paid to Belovsky's (1978) model for moose, as this is the only attempt yet to predict numerically how males and females should differ in weight. The criticisms here of nearly all the other quantitative models of body size may appear rather negative. The purpose of this chapter is not, however, solely to condemn most previous work. Evidence is presented that the energy assimilated per unit time scales at about body weight to the two-thirds power in ruminants, and it is pointed out why quantitative predictions of optimal female adult body weight from the model of Chapter 3 are not yet feasible either.

A review of existing models

Pearson (1948)

Pearson (1948) argued that in a plot of metabolic rate (measured as cubic centimetres of O_2 per hour per gram of body weight) on body weight for mammals, the extrapolated curve for shrews becomes asymptotic at about 2.5 g. Consequently, he maintained, no (adult) mammals are lighter than this, because such lighter animals would be unable to gather enough food to support their 'infinitely rapid metabolism'. This argument is unconvincing. Smaller animals also eat more, relative to their body weight (Chapter 2).

If Pearson's theory was valid, it might be predicted that smaller species should spend more time feeding. Smaller birds do feed for longer (Hinde, 1952; Gibb, 1954; Pearson, 1968). However, the interspecific dependence of feeding time on size differs between taxa. Larger primates, for example, feed for longer (Clutton-Brock & Harvey, 1977), and the same may be true for ungulates (Eltringham, 1979).

A similar argument to Pearson's was presented by Tracy (1977) who suggested that a mammal in a thermoneutral surrounding could be infinitely small, but that 'if there is a limit to the maximum heat production (metabolic rate) of a homeotherm (8), then the lower limit to size for homeotherms is also determined for particular ambient thermal environments.' Reference (8) was Pearson (1948), and Tracy simply states that Pearson pointed out that there has to be a limit to how much food can be gathered and processed to produce heat.

Kendeigh (1972); Kendeigh et al. (1977)

Kendeigh (1972) suggested that the lower limit to size in birds may be determined by the intersection of the curves for basal metabolism and existence metabolism. (Existence metabolism includes the energy expenditure on feeding, on temperature regulation and on a certain amount of movement, but not on flight.) Interspecific plots showed that in passerines these curves intersected at 2.3 g, but that in non-passerines the two curves did not intersect. No confidence limits were provided for the predicted value of 2.3 g as the lowest passerine weight, and Kendeigh *et al.* (1977) subsequently withdrew this conclusion, as when more data became available, the exponents relating existence and basal metabolism to body weight were found to be almost identical. Consequently the two curves did not intersect.

Kendeigh suggested (Kendeigh, 1972; Kendeigh *et al.*, 1977) that the upper limit to size in birds might be determined by the ability of birds to reduce their metabolism so as to tolerate high ambient temperatures. He reasoned that basal metabolic rate was the lowest rate to which metabolism could be reduced. The regression of basal metabolic rate on weight intersected that of standard metabolic rate (equal to the energy requirements of a fasting, resting bird exposed to temperatures below the zone of thermoneutrality) at 0 °C on weight at 4.6 kg in passerines, and at 16 kg in non-passerines. The largest passerine is the raven, *Corvus corax tibetanus*, which reaches 2 kg, whereas the largest extant non-passerine is the ostrich, *Struthio camelus*, which weighs about 100 kg. The fossil *Aepyornis* may have weighed 440 kg. Again, no confidence limits were provided for the values of 4.6 and 16 kg, nor is it obvious why standard metabolism at 0 °C was used as one of the two curves.

Iberall (1973)

Iberall (1973) suggested that the lower limit to body size in mammals was determined by that size at which the arteries became so narrow that viscous losses became 'appreciable'. His calculations showed that this would happen at a weight of 1.5 g. It is not apparent, however, why this problem is surmountable in juvenile mammals weighing less than 1.5 g, nor why smaller adult mammals could not adopt some other method of nutrient and oxygen transport, as used by other animals (e.g. insects), although this would perhaps be impossible for homeotherms with their high metabolic requirements (but see Chapter 9). Iberall maintained that the upper limit to mammalian size occurs when 'Greater surface stress would "pinch" or ulcerate tissues if maintained.' He calculated that this would happen at body weights in excess of 290 000 kg. The largest mammals (female blue whales, *Balaenoptera musculus*) weigh up to about 120 000 kg.

Belovsky (1978)

The only model to have made quantitative predictions for the male and female weights of any species was presented by Belovsky (1978) for moose, *Alces alces*. This important paper has been widely and favourably cited (e.g. Alexander, 1982; Krebs & Davies, 1987). Belovsky's model appears exciting as it can easily be adapted to suit other organisms. Here I first apply it to a closely related species, red deer, *Cervus elaphus*, and then critically examine Belovsky's assumptions.

Belovsky obtained an equation that related M_R, a moose's metabolism for maintenance, growth and reproduction (in kilocalories per day) to its body weight, W, and a second equation that related NE, the net energy intake by the moose, again to body weight, W. Females were then predicted to be at that body weight at which NE $-$ M_R was a maximum.

Belovsky wrote

$$\mathrm{NE} = R(KD - S)/B \qquad (6.1)$$

where R is daily rumen processing capacity, K is gross caloric content of food ingested, D is percent digestibility, S is the energetic cost of moving between food plants to crop a unit weight of food, and B is food bulkiness (=wet weight/dry weight). Belovsky computed the equation

$$R = 35\,047 \log_{10} W - 57\,993 \qquad (6.2)$$

where W is in kilograms and R is in grams wet weight per day. K was taken to be 4.2 kcal/g-dry wt for deciduous leaves, 4.8 kcal/g-dry wt for forbs and 4.1 kcal/g-dry wt for aquatics. D was measured to be 72% for deciduous leaves, 86% for forbs and 94% for aquatics. Values of S were computed and found to be negligible compared to KD. B was measured as 4.0 g-wet wt/g-dry wt for deciduous leaves, 4.4 g-wet wt/g-dry wt for forbs and 20.0 g-wet wt/g-dry wt for aquatics. Finally, M_R was assumed to be a multiple of Kleiber's (1975) interspecific formula for basal metabolism

$$M = 70W^{0.75} \qquad (6.3)$$

where M is in kilocalories per day and W is in kilograms. The multiple was taken as 2 for bulls and barren cows, and as 2.7 for cows with calves.

With these values of NE and M_R, NE $-$ M_R was maximized at $W = 285$ kg. This is close to the observed average weight of female moose equal to 330 kg. Belovsky was not sure what determined male body weight. However, he argued that although there is no reason to expect males necessarily to maximize the energy they have available for reproduction, they must have some energy available. Consequently the upper intersection of the curves relating male net energy intake and metabolic requirements as functions of body weight, must set the upper limit to male weight. These two curves intersected at a male weight of 645 kg, and the greatest recorded weight for a bull moose is 630 kg, while on average bulls weigh 450 kg.

A preliminary application of Belovsky's model to red deeer can be made assuming Equations (6.1) and (6.2) and using the following values: $D = 0.6$ (Grant & Campbell, 1978; Hill Farming Research Organisation, pers. comm.); $B = 4$; $K = 4$; $M_R = 200W^{0.75}$ (Simpson, Webster, Smith & Simpson, 1978). The greatest value of $NE - M_R$ now occurs at $W = 240$ kg, and the upper intersection of NE and M_R occurs at $W = 540$ kg. Although these values are too large (Mitchell, Staines & Welch, 1977), they represent the first predictions of optimal red deer weight and only differ from the weights of the larger continental red deer by a factor of about two, suggesting that a more detailed application of Belovsky's model to red deer might prove fruitful.

Before undertaking, however, a detailed calculation for red deer that would take into account, first, values of the energetic cost to red deer of basal metabolic rate, thermoregulation, locomotion and the other components of average daily metabolic rate, and, secondly, values for red deer of rumen size and food digestibility, caloric value and bulkiness, it is worth examining some of the assumptions of Belovsky's calculations (Reiss, 1986b).

Belovsky's equation for M_R is allometric. However, Equations (6.1) and (6.2), taken together, differ radically from the allometric form of energy assimilation as a function of body weight, namely $NE = k_2 W^b$, where k_2 and b are constants, with $b \simeq 0.67$, which most workers use and which is defended in Chapter 2. To derive Equation (6.2), Belovsky took values of the weight of food in a moose's stomach and the weight of moose from Egorov (1964), Schladweiler & Stevens (1973) and his own study. Data for the dependence of rumen size on stomach weight were unavailable for moose, so published data for white-tailed deer, *Odocoileus virginianus*, (Short, 1964) were used instead. Belovsky concluded that large moose assimilate less energy, relative to their body weight, than small moose. This conclusion rested on the observation that in white-tailed deer the rumen does not form a constant proportion of the stomach as the deer grows. This is only true, however, in white-tailed deer aged 16 weeks or less (Short, 1964)! In older animals the rumen does form a constant proportion of the stomach. In domestic cattle the corresponding figure is 6 weeks (Godfrey, 1961a, b). So if Belovsky had used data from white-tailed deer older than 16 weeks or domestic cattle older than six weeks, his equation for NE would have been proportional to $W^{1.0}$, and NE and M_R would never have intersected, except at $W = 0$. His model would have been unable to predict either adult female or adult male body weight.

Belovsky used a value for throughput time (from Mautz & Petrides, 1971) that he implicitly assumed was independent of W. In fact the speed of movement of food through the mammalian stomach is interspecifically independent of body weight, so that food takes longer to pass through the digestive system of larger species (Clements & Stevens, 1980). If throughput rate – the distance food travels per unit time in the stomach – and digestibility are independent of W, the length of the digestive system proportional to $W^{\frac{1}{3}}$, and the weight of food in the stomach proportional to $W^{1.0}$, then the volume of food assimilated per unit time should be proportional to $W^{\frac{2}{3}}$. As all herbage has approximately the same energetic value (Golley, 1961; Jordan, 1971) this would mean that NE, the energy assimilated per unit time, should be proportional to $W^{\frac{2}{3}}$. Intraspecific data on throughput time are available for the fish *Limanda limanda*, where throughput time scales as $W^{0.39}$, and stomach volume is linearly proportional to W (Jobling, Gwyther & Grove, 1977), and for the fish *Megalops cyprinoides*, where throughput time scales as $W^{0.41}$ (Pandian, 1967; Jobling *et al.*, 1977). Interspecifically, retention time in mammals scales on body weight with an exponent of 0.28 (Demment, 1983). Digestibility is approximately independent of body weight (Pandian, 1967; Arman & Hopcraft, 1975; Chapter 3), whereas the assumption that the weight of food in the stomach is intraspecifically proportional to $W^{1.0}$ seems reasonable as this is known interspecifically to be the case in insects, birds and mammals (Brody, 1945; Ledger, 1968; Hainsworth & Wolf, 1972b; Parra, 1978; Mathavan & Muthukrishan, 1980; Morton, Hinds & MacMillan, 1980; Demment, 1982; Clutton-Brock & Harvey, 1983; Demment & van Soest, 1985). Here this last assumption, that intraspecifically the weight of food in the stomach is proportional to $W^{1.0}$, is tested for red deer.

The raw data were kindly given to me by Brian Mitchell. Full background details are given in Mitchell, McCowan & Nicholson (1976). Briefly, the deer were shot on the Isle of Rhum, Inner Hebrides, Scotland, in 1969, 1970 and 1971 in February, March, April, May, July, September, October and November. Here data are analysed only from February to May inclusive, as in the other months the data are confounded by the summer abundance of food, and the autumn rut when stags (adult males) greatly reduce their feeding. Adults were from 5 to 10 years old inclusive, and were aged by a combination of tooth wear and annual dental cement layers. Milk and yeld hinds (adult females with and without calves respectively) were distinguished by a number of methods, particularly, from March to May, by the relative amounts of hair on their mammary glands. Calves were less than one year old. Animals were gutted and bled immediately after shooting and the amount of blood lost was not measured. Unpublished data collected on Rhum (cited in Mitchell *et al.*, 1976) indicate that the amount of 'bleedable blood' constitutes about 3.5% of the total weight of the animal. Accordingly, I multiplied bled body weights (including stomach contents) by 1.035 to compute live body weights. This makes the implicit assumption that intraspecifically blood weight scales on bodyweight with an exponent of 1. Interspecifically blood volume scales as $W^{0.99}$ (Günther & León de la Barra, 1966).

The data are plotted in Figure 6.1 and the results of analyses are given in Table 6.1. The correlation coefficients are similar for untransformed and logged data. *A priori* tests were made on the logged data on possible differences in slopes or intercepts between the following classes of animals:

(a) male calves versus female calves;
(b) milk hinds versus yeld hinds;

(c) adult males versus adult females;
(d) adults versus calves.

The only significant differences ($p < 0.05$) were in (c), which shows that adult males have higher slopes ($t = 3.53$, $0.001 < p < 0.01$ (least squares); $t = 2.33$, $0.02 < p < 0.05$ (reduced major axis)), and lower intercepts ($t = 3.48$, $0.001 < p < 0.01$ (least squares); $t = 2.42$, $0.02 < p < 0.05$ (reduced major axis)).

The data fail, overall, to reject the null hypothesis that the weight of food in the stomach is allometrically dependent on body weight (excluding stomach contents) with an exponent of 1, although within adult males the exponent may be greater than 1.

Another indication of the dependence on W of NE in ruminants comes from the scaling on W of rumen fermentation rates, which measure the rate of breakdown of a unit weight of the rumen contents, per unit time. Using 78 animals from 10 East African ruminant species, Hoppe (1977) found that rumen fermentation

Figure 6.1. Weight of stomach contents as a function of body weight in red deer. Symbols: □, female calves; ■, male calves; ▲, milk hinds; △, yeld hinds; ●, stags. Data from Mitchell (pers. comm.).

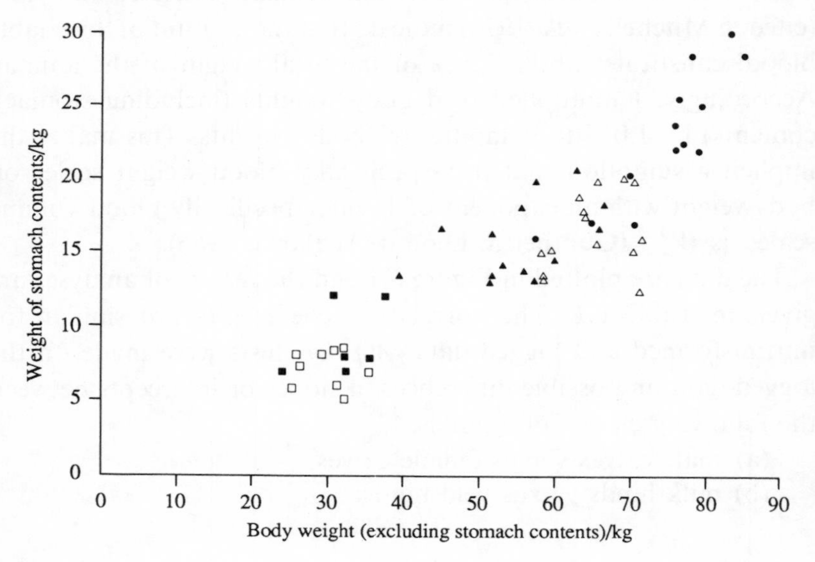

Table 6.1. *Correlation coefficients, r, and slopes of regressions of weight of stomach contents on body weight (excluding stomach contents) in red deer (data from Mitchell, pers. comm.)*

Animals	Sample size	Untransformed data	Logged data			
		r	r	least squares slope	reduced major axis slope	95% confidence limits of slopes
Male calves	6	0.446	0.465	+0.749	+1.613	±1.982
Female calves	9	0.132	0.125	−0.196	−1.561	±1.384
All calves	15	0.324	0.286	+0.502	+1.756	±1.008
Milk hinds	14	0.471	0.456	+0.505	+1.107	±0.620
Yeld hinds	14	0.354	0.356	+0.736	+2.069	±1.216
All hinds	28	0.389	0.383	+0.417	+1.089	±0.406
Stags	13	0.864	0.873	+1.674	+1.918	±0.621
All adults	41	0.762	0.718	+1.013	+1.411	±0.318
All animals	56	0.895	0.907	+1.092	+1.204	±0.138

rates scaled as $W^{-0.22}$, with 95% confidence limits of the exponent equal to ± 0.04, $r = 0.79$. This implies that NE scales as $W^{0.78}$.

Pyke (1978)

Pyke (1978) hypothesized that the bumblebee *Bombus appositus* has a body size that maximizes the bumblebee's net rate of energy intake while collecting nectar. A quantitative model predicted a body weight of 0.24 g, close to the observed value of 0.28 g. Sadly, as Pyke noted, this species is colonial and so the optimal body size of individuals from the colony's point of view (cf. Oster & Wilson, 1978) should be much less than 0.24 g, '. . . since the net rate of energy intake does not change very greatly with changing body size, the colony optimum would . . . be *much* less than the individual optimum (Smith & Fretwell, 1974). That the observed body size is close to the individual optimum suggests, therefore, that as the number of bumblebees produced by a colony increases (as a result of smaller size of individuals) the amount of time available to each individual for foraging must decrease and that the magnitude of this decrease is such that the *total* amount of *colony* foraging time is approximately independent of the size of the individual bumblebees in the colony.' This seems incredibly unlikely.

Baur & Friedl (1980)

Baur & Friedl (1980) predicted maximal and minimal body lengths for reptiles, mammals and dinosaurs. Their optimization criterion was that animals evolve so as to minimize 'mean animal activity rate' equal to the ratio of mean speed of travel to body length. They presented no evidence in support of this remarkable assumption.

Wilson (1980)

The most convincing computation for optimal body size is presented by Wilson (1980) for workers of the leaf-cutter ant, *Atta sexdens*. The main activity of these workers is to supply the colony with leaves. It transpires that the workers that are energetically the most efficient at this task have head-widths in the size range 2.2–2.4 mm, and leaf-cutting is performed mainly by workers with head-widths of 1.8–2.8 mm, although the head-widths of workers vary from 0.4 to 4.6 mm. Workers with head-widths outside the range 1.8–2.8 mm perform other tasks. Wilson's conclusion held even when the energetic cost to the colony of constructing the workers was taken into account (cf. Pyke (1978)).

Economos (1981)

Gravity was invoked as the determinant of the upper limit to body weight in mammals by Economos (1981). Apparently, small animals can adapt and survive for extended periods in stronger gravitational fields than can large animals. Extrapolations from data for three mammalian quadrupeds of the greatest gravitational fields they can tolerate gave a predicted greatest body mass (when the maximal gravitational field tolerable equals the actual gravitational field) of 20 000 kg, which 'agrees well' with the estimated mass of 20 000 kg for *Baluchitherium*, the largest mammal known to have lived on land. The reason why small mammals can exist in stronger gravitational fields than large mammals is unknown. Economos excluded man, who has a low gravity tolerance, from his regression on the grounds of bipedality.

Olive (1981)

Olive (1981) calculated that autumn-maturing temperate orb-weaving spiders should have cephalothorax widths greater than 2–3 mm. This agrees well with an observed lower limit of 3.5 mm. Olive's optimization criterion and methodology seem reasonable. Laboratory feeding experiments and insect trap samples were used to simulate orb-weaving foraging behaviour in various habitats and seasons. Spiders with cephalothorax widths less than 2–3 mm were predicted to have a greater net energy intake if they matured in the spring than if they matured in the autumn. It is difficult to comment on the details of Olive's calculations because few data are presented. However, no sensitivity analyses were performed, and one assumption seems surprising, namely that web area is independent of spider body size. As one might expect, the two are in fact strongly correlated (Olive, 1979).

Roff (1981)

Roff (1981) presented a detailed model of optimal body size in *Drosophila melanogaster*. It was assumed that individuals were selected to maximize r, the rate of increase. Roff's model predicted that *Drosophila melanogaster* should have a thorax length of about 0.95 mm. This agrees impressively with an observed range in thorax length of between 0.90 and 1.15 mm. However, as pointed out by Ricklefs (1982), Roff assumed that development time (egg to adult) was given by the equation

$$\text{Development time} = bL^{\delta} + c$$

where b, c and δ are constants and L 'some metric such as wing length or thorax length'. Using data from Robertson (1960), Roff assumed that δ equalled 3. Sensitivity analysis allowed δ to vary between 3 and 4. As Ricklefs notes, Robertson's data give various values for δ, some negative!

Ricklefs stated that values of δ 'must be determined empirically'. On theoretical grounds we expect δ to lie between approximately 0.75 and 1.0. During growth, we have (from Chapter 5):

$$\frac{\mathrm{d}W}{\mathrm{d}t} \propto W^{0.7}$$

where t is development time. Therefore

$$\int \frac{\mathrm{d}W}{W^{0.7}} \propto t$$

and so

$$t \propto W^{0.3}$$

On the assumption that $W \propto L^3$, we therefore have:

$$t \propto L^{0.9}$$

Ricklefs also pointed out that Roff chose parameter values such that the maximum value of r was about 0.26, representing a population doubling time of between 2 and 3 days. Ricklefs noted that although *Drosophila* populations may achieve such rates of increase in nature for brief periods, it is not clear that strong forces of selection are restricted primarily to periods of rapid population increase. Ricklefs calculated that for $r = 0$, a thorax length of 1.46 mm is predicted, which is 'well outside the range of observed values'. On the assumption that $\delta = 1$ and $r = 0$, the optimal thorax length for *Drosophila melanogaster*, from Roff's model, with Ricklefs' Equation (1), equals 13.52 mm! Such *Drosophila* would weigh about 2500 times more than those found in nature.

Schmidt-Nielsen (1984)

Schmidt-Nielsen (1984) and Calder (1984) provide readable summaries of ideas about factors that might set an upper or a lower limit to size in birds and mammals. The largest flying animals weigh about 12 kg (Kori bustard, *Ardeotis kori*; white pelican, *Pelacanus onocrotalus*; mute swan, *Cygnus olor*; Californian condor, *Gymnogyps californianus*). Pennycuick (1972) and Tucker (1977) point out that an upper limit to size probably exists for animals that fly. This is because the energetic requirements for flight per unit time scale at about $W^{1.0}$, whereas energy intake scales only at about $W^{0.7}$ (Chapter 2). The larger an organism, therefore, the greater the proportion of its energy intake it must apportion to flight, unless, of course, it spends less time flying, as is indeed the case in birds.

In a study of hovering moths, Casey (1981) found that the cost of hovering scaled at about $W^{1.08}$, whereas total metabolic power output scaled at about $W^{0.77}$. At a body mass of about 100 g, the two curves intersect; 100 g is about the largest body mass of birds that are able to hover for more than a few seconds. As Schmidt-Nielsen notes, hovering flight is metabolically about three times as expensive as forward flapping flight. The maximum weight for a forward-flying bird should therefore be 81 times the maximum weight of a hovering bird. (This figure is arrived at as follows. If the specific metabolic power – equal to energy requirements per unit mass – scales as $W^{-0.25}$, it takes an 81-fold increase in mass to decrease the specific power threefold.) Given 100 g as the maximum mass for an organism that hovers, the upper limit for forward flight should be about 8 kg, not far from reality.

Schmidt-Nielsen suggests that the lower limit to mass in mammals and birds may be set by the contraction time of the heart. The maximum heart rate for shrews and hummingbirds is in the range of 1200–1400 beats per minute; that is, one heartbeat lasts between 40 and 50 ms. 'In that short time, the heart must undergo a full cycle, be filled with venous blood, contract and eject the blood, relax, and be ready for the next filling cycle. It is unlikely that the design of the mammalian heart could permit it to be filled and undergo the full contraction cycle in a shorter time. If increased frequency is impossible, the only way of increasing cardiac output is to increase the heart's size and thus the stroke volume. Indeed it seems that this solution has already been necessary for both hummingbirds and shrews, which have hearts that are two or three times as large as we would expect from the general scaling of mammalian and bird hearts'.

It is worth stressing that none of these considerations pertains to why any particular species is the size that it is.

Sensitivities in body size models to changes in parameter values

Models of optimal body size should clearly state and defend the optimization criterion adopted, carefully list the constraint equations and subject the quantitative conclusions to thorough sensitivity analyses.

Let us suppose, as in earlier chapters, that a free-living adult female's energy requirements per unit time for everything except reproduction, E_{req}, can intraspecifically be modelled as

$$E_{req} = k_1 W^a$$

whereas intraspecifically E_{in}, her energy assimilation per unit time, can be modelled as

$$E_{in} = k_2 W^b$$

Then the energy available for reproduction per unit time, E_{rep}, is given by

$$E_{rep} = E_{in} - E_{req} = k_2 W^b - k_1 W^a$$

On the assumption that adult female body weight, W_f, has evolved so as intraspecifically to maximize E_{rep}, we have

$$\frac{d}{dW}(k_2 W^b - k_1 W^a) = 0$$

So

$$W_f^{a-b} = k_2 b / k_1 a \qquad (6.4)$$

It can now be seen that precise quantitative predictions of W_f are likely to prove extremely difficult. From Equation (6.4) a 20% increase in k_2/k_1 leads to a 1000% increase in W_f! Consequently slight errors in the measurement of k_1/k_2 lead to huge errors in the prediction of W_f.

A second difficulty arises if lifespan depends intraspecifically on body weight. Suppose that, as muted in Chapter 5, lifespan scales intraspecifically as W^{1-a}, and that females are selected to maximize lifetime reproductive success, which I will equate with lifetime parental investment (measured in units of joules), LPI. Then we have

$$\text{LPI} \propto W^{1-a}(k_2 W^b - k_1 W^a) \qquad (6.5)$$

$$\frac{d}{dW}\text{LPI} = 0 \qquad (6.6a)$$

$$\frac{d^2}{dW^2}\text{LPI}|_{W=W_{fl}} < 0 \qquad (6.6b)$$

where W_{fl} is that adult female body weight at which lifetime parental investment is maximized. W_{fl} does not equal W_f. We have

$$(W_{fl}/W_f) = (1 + b - a)a/b \qquad (6.7)$$

As one might expect, W_{fl} exceeds W_f, provided that a is greater than b. This is because larger animals live longer. For $a = 0.75$, $b = 0.67$, we have

$$W_{fl}/W_f = 1.45 \qquad (6.8)$$

Conclusions

A review is made of quantitative models of body size. Practically nothing is known of why any species is the size it is, with the exception of Wilson's (1980) study on workers of the leaf-cutter ant, *Atta sexdens*. Models of optimal body size should clearly state and defend the optimization criterion adopted, carefully list the constraint equations and subject any conclusions to thorough sensitivity analyses.

Arguments are given for expecting the energy assimilated per unit time to scale at about body weight to the two-thirds power in ruminants.

A reason is suggested for why accurate predictions of body size are apparently so difficult. It seems that even very small errors in the determination of a key variable, k_1/k_2, lead to massive errors in the prediction of body weight.

Finally, the consequence to optimal body size models of females being selected to maximize lifetime rather than annual parental investment are investigated. This difference will be of importance if lifespan depends intraspecifically on size.

7

Sexual dimorphism in body size

Many authors, dating back to Darwin (1871), have considered factors that might affect the degree of sexual dimorphism (reviewed by Clutton-Brock, Harvey & Rudder, 1977; Mace, 1979). Frequently, however, only one sex is considered, or it is unclear what limits the degree of sexual dimorphism. For example, intermale conflict is believed to select for increased male size (Darwin, 1871; Selander, 1972; Trivers, 1972). However, larger females achieve greater reproductive success (Chapter 4) and Ralls (1976a) has argued that this may be why females are sometimes larger than males. On the other hand, Erlinge (1979) and Moors (1980) have argued that smaller females enjoy a bioenergetic advantage.

It is not yet clear what sets an upper limit to male size in species where males compete over females, and larger males enjoy a competitive advantage. One possibility is male-biased mortality. In many species where males are the larger sex, mortality is greater for males than for females (Potts, 1970; Selander, 1972; Clutton-Brock & Albon, 1982). Selander (1972) argued that the greater mortality of male great-tailed grackles, *Quiscalus mexicanus*, occurs partly because males are more vulnerable to predators, and partly because males are above the optimal size for efficient foraging. Another possibility, suggested by Belovsky (1978) and Searcy (1979), is that larger males have less energy available for reproduction.

The greatest degree of sexual dimorphism, in species where males are heavier than females, occurs in pinnipeds (Scheffer & Wilke, 1953; Bryden, 1969). Bartholomew (1970) suggested three factors which might set an upper limit on male size in pinnipeds. First, larger males are more aggressive and in the most dimorphic species oblivious of the pups, so that many pups are squashed to death. 'To the extent that males may inadvertently kill their own progeny, this situation obviously reduces the reproductive effectiveness of their aggressive behaviour'. Secondly, 'Dominant male elephant seals will sometimes interrupt a copulation to chase away an encroaching male'. Thirdly, male elephant seals 'are clearly near a size beyond which they could not adequately defend a terrestrial territory, because of their limited locomotor ability on land.'

A quantitative theory of sexual dimorphism is clearly desirable. The only one provided so far is by Belovsky (1978). This applied to just one species, moose, and was criticized in Chapter 6. In this chapter the intraspecific dependence of male reproductive success on body weight is first investigated. This is then included in a quantitative model of the degree of sexual dimorphism in body weight. Finally, the amount of energy available to males for reproduction is considered.

Male body weight and reproductive success

Data from eight studies are analysed here. Two forms of the relationship between male body weight and reproductive success are considered: a linear and an allometric dependence of reproductive success on body weight. Accordingly, least squares regressions are calculated for reproductive success on body weight and log reproductive success on log body weight. The correlation coefficients are listed in Table 7.1. The data are either lumped or unlumped. Lumped data include all individuals, but aggregated into weight classes. In the case of unlumped data the individuals are not aggregated into weight classes. In three of the studies, unlumped data were not available. Individuals with reproductive success equal to zero were excluded from the analyses of the unlumped data as $\log 0 = \infty$. Pertinent information on each study is listed immediately below.

Table 7.1. *Correlation coefficients, r, for intraspecific relationships between male weight and reproductive success*

Species	Sample size*	Regression of reproductive success on weight		Regression of log reproductive success on log weight		Source
		Unlumped data	Lumped data	Unlumped data	Lumped data	
Anolis garmani	523, —, 9	—	+0.973	—	+0.915	(1)
Papio anubis	15, 11, 5	+0.252	+0.793	+0.123	+0.890	(2)
Bufo bufo	428, —, 4	—	+0.993	—	+0.992	(3)
Rana catesbeiana 2°	37, 20, 5	+0.235	+0.997	+0.270	+0.849	(4)
Rana catesbeiana 3°	37, 20, 5	+0.347	+0.969	+0.366	+0.956	(4)
Rana catesbeiana 4°	37, 20, 5	+0.527	+0.966	+0.611	+0.989	(4)
Bison bison	17, 12, 5	−0.268	+0.102	−0.247	+0.143	(5)
Rana sylvatica	346, —, 7	—	+0.967	—	+0.954	(6)
Bufo calamita	21, 18, 5	+0.345	+0.866	+0.353	+0.895	(7)
Cervus elaphus	26, 26, 5	+0.392	+0.674	+0.513	+0.547	(8)

Sources: (1) Trivers, 1976; (2) Popp, 1978; (3) Davies & Halliday, 1979; (4) Howard, 1979; (5) Lott, 1979; (6) Howard, 1980; (7) Arak, pers. comm.; (8) Guinness *et al.*, pers. comm.

* Sample size (n_1, n_2, n_3). n_1 = number of individuals; n_2 = number of individuals with reproductive success $\neq 0$; n_3 = number of lumped classes.

Anolis garmani, lizard, Trivers (1976):
Reproductive success measured as the frequency of copulation. Data collected from July 1969 to January 1972 for a total of eight months. I assumed weight proportional to $(\text{length})^3$.

Rapio anubis, olive baboon, Popp (1978):
Reproductive success measured as the sum of relative consort success (equal to the proportion of time spent in consort with females) and relative copulation success (equal to the number of copulations achieved per unit time each male was seen). Data collected from May 1976 to March 1978 for a total of nine months. Weights given.

Bufo bufo, European common toad, Davies & Halliday (1979):
Reproductive success measured as the probability of spawning. Observations spanned the 1978 breeding season. I assumed weight proportional to $(\text{length})^3$.

Rana catesbeiana, bullfrog, Howard (1979):
Three measures of reproductive success used. 2° reproductive success equals the number of copulations achieved; 3° reproductive success equals the number of zygotes sired; 4° reproductive success equals the number of zygotes that survive to hatching. Observations spanned the 1976 breeding season. I assumed weight proportional to $(\text{length})^3$.

Bison bison, American bison, Lott (1979):
Reproductive success measured as the number of copulations. Observations spanned the 1972 breeding season. Weights given.

Rana sylvatica, woodfrog, Howard (1980):
Reproductive success measured as the number of matings. Observations spanned the 1976 breeding season. I assumed weight proportional to $(\text{length})^3$.

Bufo calamita, natterjack toad, Arak (pers. comm.):
Reproductive success measured as the number of matings per day at the pool. Observations spanned the 1980 breeding season. Weights given.

Cervus elaphus, red deer, Guinness, Clutton-Brock & Albon (pers. comm.):

Reproductive success measured as the number of hind-days held. This equals the number of days that a stag held a hind in his harem, summed over all hinds for each stag. This measure of reproductive success is defended by Gibson (1978) and Gibson & Guinness (1980). Observations spanned the four breeding seasons 1974 to 1977 inclusive. I calculated body weights from antler weights, re-analysing the data in Huxley (1932) for stags weighing less than 200 kg to determine the allometric dependence of antler weight on body weight. Least squares regression was used: $r = 0.996$. The resultant body weights are rather low (Appleby, pers. comm.). Presumably Huxley used some measure of carcase weight rather than live body weight. The unlumped data on body weight and reproductive success are plotted in Figure 7.1.

Table 7.1 shows that the linear and the allometric functions of body weight correlate equally well with reproductive success.

Figure 7.1. Reproductive success as a function of body weight in male red deer. Data from Guinness *et al.* (pers. comm.).

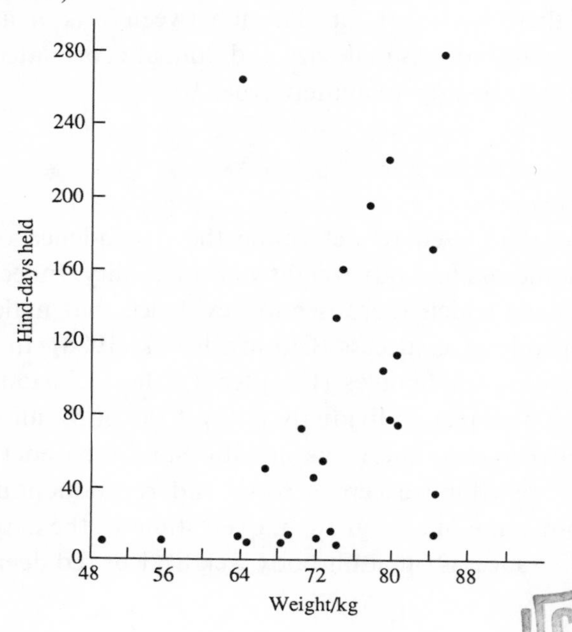

One reason why male reproductive success might be allometrically related to body weight is that in many species there is assortative mating for size (reviewed by Ridley & Thompson, 1979; Greenspan, 1980; Brown, 1981) and female reproductive success is allometrically related to female body weight (Chapter 4).

A second reason why male reproductive success might be allometrically related to body weight is that in many species larger males are more likely to be dominant (Potter, Wrensch & Johnston, 1976; Davies & Halliday, 1978; Clutton-Brock *et al.*, 1979; Veuille & Rouault, 1980) and dominant males achieve disproportionately greater reproductive success by securing access to large numbers of females (Nozawa, 1960; Le Boeuf, 1974; Gibson & Guinness, 1980). Suppose that offspring are arranged geometrically among males so that the dominant male sires a proportion p, the second dominant a proportion $p(1 - p)$, and the 'n'th male a proportion $p(1 - p)^{n-1}$. Then the condition for male reproductive success to be allometrically related to body weight is for

$$p(1 - p)^{n-1} \propto W_{\mathrm{m}}^{c} \tag{7.1}$$

where W_{m} is male body weight, and c is the slope of the regression of log reproductive success on log body weight. Equation (7.1) is satisfied if there is a linear correlation between rank, n, and log W_{m}. Because in many species male size and rank are correlated (*loc. cit.*), this is likely to be approximately true.

Discussion

The data used to determine the dependence of male reproductive success on body weight vary in quality. Indeed there are few species for which there is good evidence that males differ in lifetime reproductive success (Clutton-Brock, 1988). In particular, as was the case for females (Chapter 4), the inclusion in the regressions of younger individuals not yet devoting all of the difference between their energy assimilation and their energy requirements for everything except growth and reproduction to reproduction, but some of it to growth, overestimates the dependence of reproductive success on adult body weight. For red deer and bison ages were known and sub-adults were excluded from the regressions.

A model of sexual dimorphism in body size

As before, we have that for adult females the energy available for reproduction per unit time, E_{rep}, is given by

$$E_{rep} = E_{in} - E_{req} = k_2 W^b - k_1 W^a$$

For the females of many species the energy available for reproduction, E_{rep}, is limiting (McNab, 1963b; Klein, 1964; Watson, 1970; Sadleir, Casperson & Harling, 1973; Randolph *et al.*, 1977; Sinclair, 1977; Mitchell *et al.*, 1977; Belovsky, 1978). In particular Belovsky (1978) on moose, *Alces alces*, and Searcy (1979) on red-winged blackbirds, *Agelaius phoeniceus*, suggest that the energy available for reproduction is size-dependent, and that adult breeding females have adjusted their body size so as to maximize it.

If we now assume, as mentioned in Chapters 5 and 6, that adult female body weight, W_f, has evolved so as intraspecifically to maximize E_{rep}, we have

$$\frac{d}{dW}(k_2 W^b - k_1 W^a) = 0 \qquad (7.2)$$

and

$$\frac{d^2}{dW^2}(k_2 W^b - k_1 W^a)|_{W=W_f} < 0 \qquad (7.3)$$

So

$$W_f^{a-b} = k_2 b / k_1 a \qquad (7.4)$$

(Sebens, 1979), and

$$a > b \qquad (7.5)$$

These calculations assume that just the energy available for reproduction is limiting. In some species, however, other resources such as protein, specific amino acids, sodium or other nutrients may also or instead be limiting (Blair-West, Coghlan, Denton, Nelson, Orchard, Scoggins, Wright, Myers & Junqueira, 1968; Randolph *et al.*, 1975; Harper, 1977; Belovsky, 1978; Altmann & Wagner, 1978; White, 1978; Greenstone, 1979). In fact the allometric form of the model and its conclusions quite possibly hold for any limiting resource. For example, nitrogen requirements scale as $W^{0.72}$ and sulphur requirements as $W^{0.74}$ (Brody, 1945), whereas protein uptake and requirements probably depend on body size in the same way that energy intake and requirements do (Moen, 1973). Practically all the relevant data, however, relate to energy, and so energy-limitation is considered throughout.

Other workers have suggested that body size is energetically constrained (Schoener, 1969; Kendeigh, 1972; Selander, 1972; Downhower, 1976; Erlinge, 1979; Searcy, 1980) and the model presented here initially assumes that female body weight is determined solely by energetic considerations. This is bound to be a simplification, since other factors are known to affect body size (Pyke, 1978).

The analysis leading to Equation (7.4) only considered females. What do males maximize?

Monogamous species are typically characterized by high parental investment (Wilson, 1975; Kleiman, 1977; Maynard Smith, 1977). In such species we might expect adult males as well as females to maximize the difference between their energy assimilation and non-reproductive energy requirements, that is to maximize E_{rep}. Consequently W_m should equal W_f and the degree of sexual dimorphism in body weight should equal 1. In monogamous species W_m/W_f is, perhaps unsurprisingly, close to 1 (Ralls, 1976a; Clutton-Brock *et al.*, 1977; Alexander *et al.*, 1979).

For species where males cannot simply be expected to maximize E_{rep}, other assumptions can be made.

Initially assume that male reproductive success is proportional to the amount of time a male spends trying to reproduce (e.g. looking for receptive females), and that while trying to reproduce a male can assimilate no energy. Then if t is the proportion of time spent assimilating energy and $(1 - t)$ is the proportion of time spent trying to reproduce, we have

$$m = k_4(1 - t) \quad (7.6)$$

where m is male reproductive success and k_4 is a constant.

Next assume that for males, as for females, energy assimilation per unit time spent assimilating energy is $k_2 W^b$, and that non-reproductive energy requirements are $k_1 W^a$ per unit time not spent trying to reproduce. Finally, assume that a male uses energy at a rate equal to $k_3 W^a$ per unit time spent trying to reproduce. Now, for a male we have that the energy available for reproduction, E_{rep}, is given by

$$E_{rep} = (k_2 W^b - k_1 W^a)t \tag{7.7}$$

while the energy used for reproduction equals

$$k_3 W_m^a(1 - t) \tag{7.8}$$

Equating Equation (7.7) with Equation (7.8) and substituting for t into Equation (7.6) gives

$$m = k_4 \frac{1 - \dfrac{k_1}{k_2} W_m^{a-b}}{1 - \left(\dfrac{k_1 - k_3}{k_2}\right) W_m^{a-b}} \tag{7.9}$$

Because k_1, k_2 and $k_3 > 0$, Equation (7.9) has its maximum at $W_m = 0$. This is because the exponent of energy intake on body weight is lower than the exponent of non-reproductive energy requirements on body weight (Equation (7.5) and Chapter 5). Consequently, smaller males need to spend less time feeding, and so are able to spend more time trying to reproduce. As male reproductive success was assumed proportional to the length of time a male spends trying to reproduce (Equation (7.6)), the optimal weight is predicted to be zero.

Clearly this prediction is unrealistic – at the very least males need $W_m \neq 0$ so as to produce sperm. Larger males often achieve greater reproductive success, as reviewed earlier. Two ways to incorporate the fact that reproductive success depends on male size as well as on the time males can spend trying to reproduce are to write

$$m = k_4(1 - t)W_m^c \tag{7.10}$$

or

$$m = (k_5 W_m + k_6)(1 - t) \tag{7.11}$$

where k_4, k_5 and k_6 are constants.

For the eight species considered earlier, the allometric (Equation (7.10)) and the linear (Equation (7.11)) functions of body weight correlate equally well with reproductive success. Here the allometric function is preferred for three reasons. First, it has the advantage, unlike Equation (7.11), that $m = 0$ for $W_m = 0$. Secondly, the combination of Equations (7.7), (7.8) and (7.11) leads to an equation in W_m that is analytically insoluble even with the simplification $k_1 = k_3$. Thirdly, as considered earlier, for some species the allometric dependence of reproductive success on male body weight is expected.

Equating Equation (7.7) with Equation (7.8) and substituting for t into Equation (7.10) gives

$$m = k_4 \frac{k_2 W^c - k_1 W^{a-b+c}}{k_2 + (k_3 - k_1) W^{a-b}} \tag{7.12}$$

To find the value of W, W_m, which maximizes Equation (7.12), Equation (7.12) is differentiated with respect to W and set equal to 0. Substituting then for k_2 from Equation (7.4), we arrive at

$$\left(\frac{W_m}{W_f}\right)^{a-b} = \frac{a}{2b(K-1)}\left(\frac{(b-a)K}{c} + (K-2) + \left[\left(\frac{(b-a)K}{c} + (K-2)\right)^2 + 4(K-1)\right]^{\frac{1}{2}}\right) \tag{7.13}$$

where $K = k_3/k_1$.

At first sight it might appear that Equation (7.13) is singularly uninformative. However, it does make some precise predictions. As K decreases – that is, as the cost to a male of his reproductive energy requirements relative to the cost of his non-reproductive energy requirements decreases – W_m/W_f increases. In other words, as might be expected, when males need less energy for reproduction, that is as K decreases, males can afford to be larger, so that the predicted degree of sexual dimorphism increases.

For males the energy they can invest in reproduction, E_{rep}, will equal zero when $E_{in} = E_{req}$, i.e.

$$k_2 W_m^b = k_1 W_m^a \tag{7.14}$$

There are two solutions to Equation (7.14). Either

$$W_m = 0 \tag{7.15}$$

or

$$W_m^{a-b} = k_2/k_1 \quad (7.16)$$

Combining Equations (7.15) and (7.16) with Equation (7.4), we have two extreme values of W_m/W_f. We find that either

$$W_m/W_f = 0 \quad (7.17)$$

or

$$(W_m/W_f)^{a-b} = (k_2/k_1)\cdot(k_1 a/k_2 b) = a/b \quad (7.18)$$

One way to understand the reason for Equation (7.18) is to see that there exists a male weight at which males have to spend all their time assimilating energy and have no time left for reproduction. Similarly, in Equation (7.13), if we let $K \to 0$, $(W_m/W_f)^{a-b} \to a/b$. In other words, as the cost to males of spending time trying to reproduce becomes negligible, males can afford to become larger and larger. Finally, in Equation (7.13), as we let $c \to \infty$, $(W_m/W_f)^{a-b} \to a/b$.

Consequently, for any species we have the inequality

$$0 \leqslant W_m/W_f \leqslant (a/b)^{1/(a-b)} \quad (7.19)$$

(Reiss, 1982b).

As $b \to a$, $(a/b)^{1/(a-b)} \to e^{1/a}$, where e = 2.71828 . . . . The values of $(a/b)^{1/(a-b)}$ for some values of a and b are given in Table 7.2.

The predicted values of W_m/W_f, from Equation (7.13), for some values of c and K, assuming $a = \frac{3}{4}$, $b = \frac{2}{3}$, are given in Table 7.3.

Table 7.2. *Predicted maximal values of the degree of sexual dimorphism in body weight, W_m/W_f, in species where males are heavier than females, as functions of a, the intraspecific slope of log average daily metabolic rate on log body weight, and b, the intraspecific slope of log energy assimilation on log body weight*

		a		
		0.5	0.75	1.0
b	0.333	11.40	7.01	5.20
	0.667	—	4.11	3.37

Note: These values of the maximal degree of sexual dimorphism in body weight, in species where males are heavier than females, are calculated from Equation (7.18).

Table 7.3. *Predicted values of the degree of sexual dimorphism in body weight,* W_m/W_f*, as functions of* c*, the intraspecific slope for males of log reproductive success on log body weight, and* K*, the ratio of* k_3 *to* k_1

		K		
	0.5	1.0	2.0	4.0
0	0	0	0	0
0.5	0.79	0.65	0.56	0.51
1.0	1.67	1.57	1.51	1.48
c 2.0	2.56	2.52	2.49	2.48
4.0	3.22	3.21	3.20	3.20
8.0	3.63	3.63	3.63	3.63
∞	4.11	4.11	4.11	4.11

Note: These values of the degree of sexual dimorphism in body weight are calculated from Equation (7.13), assuming $a = \frac{3}{4}$, $b = \frac{2}{3}$.

Evidently, from Table 7.3, W_m/W_f is predicted to depend more on c than on K. For the simplest case, $K = 1$, Equation (7.13) reduces to

$$(W_m/W_f)^{a-b} = ac/b(a - b + c) \qquad (7.20)$$

Values of c can be calculated for the eight species in Table 7.1. Because the measurements of weight are probably considerably more accurate than the measurements of reproductive success, least squares regression is probably more appropriate than reduced major axis regression. Calculated values of c and c' $(=c/r)$, and 95% confidence limits of c and c' for unlumped and lumped data are given in Table 7.4. Equation (7.20) predicts that, as might be expected, W_m/W_f should be greater for species with larger values of c (or c'). However, none of the Spearman rank correlation coefficients of the actual values of W_m/W_f with c or c' is significant:

r_s of W_m/W_f with c from unlumped data $= -0.60$, $p > 0.10$

r_s of W_m/W_f with c' from unlumped data $= +0.10$, $p > 0.10$

r_s of W_m/W_f with c from lumped data $= -0.46$, $p > 0.10$

r_s of W_m/W_f with c' from lumped data $= -0.01$, $p > 0.10$

(The data used for *R. catesbeiana* are the 4° estimates of reproductive success, as these are probably the best (Howard, 1979).)

Table 7.4. *Male weight and reproductive success: values of c and c'*

Species	Actual degree of sexual dimorphism in body weight	Unlumped data			Lumped data		
		c	c'	95% confidence limits	c	c'	95% confidence limits
Rana sylvatica	0.67	—	—	—	+3.524	+3.693	±1.271
Bufo bufo	0.86	—	—	—	+3.026	+3.049	±1.140
Rana catesbeiana 2°	0.86	+0.506	+1.875	±0.894	+2.577	+3.035	±2.947
Rana catesbeiana 3°	0.86	+1.256	+3.433	±1.582	+3.946	+4.128	±2.228
Rana catesbeiana 4°	0.86	+3.430	+5.617	±2.203	+4.763	+4.814	±1.288
Bufo calamita	0.99	+1.600	+4.534	±2.248	+5.036	+5.624	±4.599
Cervus elaphus	1.45	+4.733	+9.228	±3.324	+2.099	+3.837	±5.902
Bison bison	1.82	−2.468	−10.002	±6.831	+1.053	+7.361	±13.384
Papio anubis	1.92	+0.846	+6.682	±5.134	+3.050	+3.426	±2.868
Anolis garmani	2.25	—	—	—	+2.121	+2.317	±0.906

Table 7.5. *Sexual dimorphism in body weight: data and predictions*

Species	Actual degree of sexual dimorphism in body weight	Predictions from unlumped data			Predictions from lumped data		
		Lower limit	Mean	Upper limit	Lower limit	Mean	Upper limit
Rana sylvatica	0.67	—	—	—	2.66	3.10	3.34
Bufo bufo	0.86	—	—	—	2.45	2.97	3.24
Rana catesbeiana 2°	0.86	0.00	0.66	2.05	0.00	2.81	3.43
Rana catesbeiana 3°	0.86	0.00	1.90	2.90	2.33	3.20	3.50
Rana catesbeiana 4°	0.86	1.87	3.08	3.45	3.09	3.34	3.49
Bufo calamita	0.99	0.00	2.23	3.19	0.51	3.38	3.71
Cervus elaphus	1.45	2.06	3.33	3.63	0.00	2.58	3.63
Bison bison	1.82	0.00	0.00	3.28	0.00	1.65	3.84
Papio anabis	1.92	0.00	1.33	3.48	0.04	2.97	3.48
Anolis garmani	2.25	—	—	—	1.85	2.59	2.97

Note: The mean, upper and lower limits are calculated from the mean and 95% confidence limits of c in Table 7.4.

Predicted values of W_m/W_f, calculated from Equation (7.20), are listed in Table 7.5 with values of c from Table 7.4, assuming $a = \frac{3}{4}$, $b = \frac{2}{3}$.

From Equation (7.20), the condition for $W_m = W_f$ is for $c = b$. In species where individuals mate assortatively for size, male reproductive success and female reproductive success should scale on body weight with very nearly the same exponents, assuming that the frequency of reproduction is independent of size, and provided that the assortative mating is what might be termed 'linear assortative mating', namely that the ratio of male to female weight is a constant for all pairs. In such species W_m/W_f is predicted to lie close to 1, and to exceed W_m/W_f in species where there is no assortative mating for size.

Crump (1974) found no assortative mating for size in four species of anurans, and in each case males were smaller than females: *Eleutherodactylus altmazonicus*, $W_m/W_f = 0.28$; *E. variabilis*, $W_m/W_f = 0.32$; *E. croceoinguinis*, $W_m/W_f = 0.41$; *Hyla garbei*, $W_m/W_f = 0.59$, assuming in each case that weight $\propto$ (length)^3. W_m/W_f does seem to be greater, and fairly close to 1, in species with assortative mating for size: *Bufo americanus*, toad, $W_m/W_f = 0.71$ (Licht, 1976); *Chauliognathus pennsylvanicus*, soldier beetle, $W_m/W_f = 0.79$ (McCauley & Wade, 1978): *Homo sapiens*, humans, $W_m/W_f = 1.25$ (Roberts, 1977; Alexander, Hoogland, Howard, Noonan & Sherman, 1979); *Uca rapax*, fiddler crab, $W_m/W_f = 1.49$ (Greenspan, 1980); *Thermosphaeroma thermophilum*, Socorro isopod, $W_m/W_f = 1.50$ (Shuster, 1981). However, species that show assortative mating for size do not always have larger values of W_m/W_f than species that mate randomly for size. For example, in the dung fly, *Scatophaga stercoraria*, there is no assortative mating for size, yet $W_m/W_f = 1.91$ (Borgia, 1981). In this species the frequency of male reproduction is not independent of size: larger males are more successful at controlling or gaining access to oviposition sites, and are more successful in fights over females (Parker, 1970; Borgia, 1979). Similarly, assortative mating for size is probably absent or negligible in polygynous highly dimorphic ungulates, pinnipeds and primates.

Discussion

The assumption that male reproductive success is proportional to the time a male spends trying to reproduce holds for grey seals, *Halichoerus grypus*, (Anderson, Burton & Summers, 1975; Boness & James, 1979); red deer (Gibson, 1978) and various anurans (Gatz, 1981; Woodward, 1982; Arak, 1983; Ryan, 1983). It is perhaps unlikely to hold well for species that breed for only a very short time each year, for example those anurans that have an annual breeding season of only a day or two (Wells, 1977) and the praying mantis, *Acanthops falcata*, which breeds only at first light (Robinson & Robinson, 1979). Sullivan (1987) found, via stepwise multiple regression analyses, that there was a weakly positive but non-significant relationship between the number of nights a male spent at a chorus and mating success in Woodhouse's toad, *Bufo woodhousei*.

One could imagine across species that the relationship between breeding season length and the ratio of the variance in male re-

Figure 7.2. Hypothetical relationship between the ratio of the variance of male to female reproductive success as a function of the length of the breeding season.

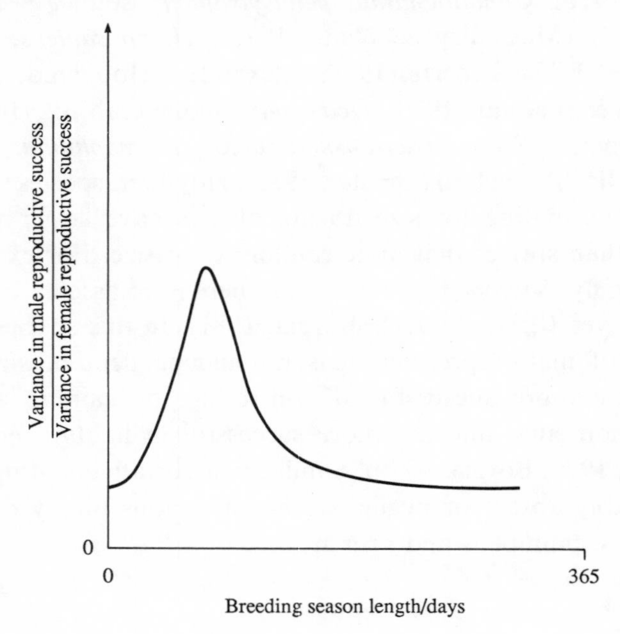

productive success to the variance in female reproductive success might be as suggested in Figure 7.2. For species in which all reproduction is squashed into a very short period each year, extreme scramble competition will mean that the ratio of male to female variance in reproductive success is relatively low. As the breeding season lengthens, there exists the potential for a handful of males to dominate breeding. Eventually, however, the breeding season lasts for so long that each male can only devote his energies towards reproduction for a small proportion of the entire breeding season.

The assumption that males cannot assimilate energy while trying to reproduce is reasonable for land-breeding seals and some other seasonal breeders such as many anurans (Wells, 1977) and many ungulates (Leuthold, 1977). In many species, males decrease their energy assimilation while trying to reproduce, but not to zero. Male chimpanzees, *Pan troglodytes*, for example, feed less when in the presence of an oestrous female (Wrangam & Smuts, 1980), but do not cease to feed. This does not affect the predicted upper limit of male size, but does mean that below this limit Equation (7.13) will underestimate the degree of sexual dimorphism in body weight.

It would be more realistic to have $(d - t)$ as the proportion of the time males spend trying to reproduce, rather than $(1 - t)$, where t is again the proportion of time spent assimilating energy and $0 \leqslant d \leqslant 1$. This still assumes that males assimilate no energy while trying to reproduce, but allows males a proportion of the time equal to $(1 - d)$ neither to assimilate energy nor to try to reproduce. This time might, for instance, be spent asleep. With this change from $(1 - t)$ to $(d - t)$, the general equation for W_m/W_f as a function of d, K, a, b and c is now analytically insoluble. A reasonable simplification is to put $K = 1$, as it was previously shown (Table 7.3) that predicted values of W_m/W_f are relatively insensitive to variations in K. When this simplification is made, the equation for W_m/W_f is the same as before, i.e. Equation (7.20).

The predicted values of W_m/W_f are very sensitive to sex-specific differences in the ratio k_2/k_1. If k_2/k_1 for the male is 10% greater than k_2/k_1 for the female of a species, W_m/W_f, for $K = 0$, or for $c = \infty$, is predicted from Equation (7.18), for $a = \frac{3}{4}$, $b = \frac{2}{3}$, to equal not 4.1 but 12.8. If k_2/k_1 for the female is 10% greater than k_2/k_1 for the male, W_m/W_f is predicted to equal not 4.1 but 1.3.

Sex-specific differences in k_2/k_1 are almost bound to exist. What is as yet unknown is the precise magnitude of these differences. A significant difference between the sexes in k_1 was looked for and not found in the weasel, *Mustela nivalis*, (Moors, 1977). An indication, however, that sex-specific differences in k_2/k_1 may exist comes from the observation that in many dimorphic species males and females differ in growth rates, for instance in red deer, *Cervus elaphus*, (Blaxter, Kay, Sharman, Cunningham & Hamilton, 1974), in humans (Tanner, 1978) and in the lizards *Anolis garmani* (Trivers, 1976) and *Basiliscus basiliscus* (Van Devender, 1978).

If k_2/k_1 has the same value for males and females within a species, W_m/W_f is predicted from Table 7.2 to lie between 0 and about 11 (*cf.* Reiss (1982b)). As predicted, the most dimorphic species known are those in which females far outweigh males so that W_m/W_f lies close to 0. Species of *Bonellia* in the phylum Echiura have extremely low values of W_m/W_f. In *B. viridis* females are hundreds of times longer than males (up to 100 cm for females, up to 3 mm for males (Barnes, 1974)) and therefore possibly millions of times heavier. Very low values of W_m/W_f also occur in some solitary haplodiploids (Hamilton, 1967), in ceratioid anglerfish (Pietsch, 1975, 1979), in several molluscs (Morton, 1981) and in many social insects. Sexual dimorphism in body weight, as predicted by Equation (7.19), is less extreme when males outweigh females. The greatest value of W_m/W_f seems to be about 8 in the southern elephant seal, *Mirounga leonina*, (Bryden, 1969). In birds the greatest values of W_m/W_f lie between 2 and 3 (Selander, 1972; Ralls, 1976b).

The predictions of W_m/W_f for the eight species in Table 7.5 are not very successful. This may be because the assumptions leading to Equation (7.20) are unrealistic. Alternatively, the values of c used in Equation (7.20) for the eight species may be inaccurate. Certainly the confidence limits of c are wide (Table 7.4) and the considerable differences between the estimates of c for the 2°, 3° and 4° data of the bullfrog illustrate the difficulties in obtaining accurate estimates of c.

The data used to calculate c in Table 7.2 are actually, with the exception of the data for the natterjack toad, of the form

$$m \propto (1 - t)W_m^c \qquad (7.21)$$

rather than

$$m \propto W_m^c$$

As t and c are not independent, it is therefore strictly invalid to estimate c from these data. The error, however, is small. Substituting for t from Equations (7.7) and (7.8) into Equation (7.21), we have

$$m \propto (k_2 W_m^c - k_1 W_m^{a-b+c})/(k_2 + (k_3 - k_1)W_m^{a-b}) \quad (7.22)$$

For $c \gg (a - b)$ this is approximately of the form $m \propto W_m^c$.

Rather little is known about the value of K, the ratio for a male of the energetic cost of a unit of time spent trying to reproduce to a unit of time spent assimilating energy. I suspect that greater values of K are found in poikilotherms than in homeotherms. Bucher, Ryan & Bartholomew (1982) report a value of 5.7 for K for the frog *Physalaemus pustulosus*. For grey seals, Anderson & Fedak (1985) report a value for K of 3.0.

In Chapter 6 it was pointed out that if natural selection acts to maximize lifetime reproductive success, rather than the rate of reproduction, a different value of adult female body weight is predicted if lifespan depends on size. If male and female lifespans are proportional to W^{1-a} we need to replace c in Equation (7.12) by $1 - a + c$. On the assumption that $K = 1$, the new equation for W_m/W_f is:

$$(W_m/W_f)^{a-b} = (1 - a + c)/(1 - b + c)(1 + b - a) \quad (7.23)$$

For $a = \frac{3}{4}$, $b = \frac{2}{3}$, $c = \infty$, Equation (7.23) predicts $W_m/W_f = 2.84$, which is lower than the previous calculation of 4.11 (Table 7.3). This is as expected, because W_m cannot exceed that value at which $E_{rep} = 0$, whereas W_f would be greater (Equation (6.8)). The value of c necessary for $W_m = W_f$ is, from Equation 7.23, still $c = b$. It is premature to assert definitely that lifespan scales intraspecifically as W^{1-a}. Indeed, Geist (1971) provided evidence that smaller male mountain sheep live longer, and this is also the case in the woodlouse *Asellus aquaticus* (Ridley & Thompson, 1979). In species where larger males fight more, or are exposed to greater risks of starvation or predation, it may often be the case that smaller males live longer.

From Equations (7.20) and (7.23) the condition for $W_m = W_f$ is for $c = b$. This may be the explanation for an apparent paradox, namely that in some species $W_m < W_f$, yet larger males are known to do better in contests for access to females. This is the case in *Drosophila melanogaster* (Partridge, Hoffman & Jones, 1987), in the European common toad (Davies & Halliday, 1977, 1978, 1979) and some other anurans (Wells, 1977; Howard, 1978a, b, 1979, 1980). In the Weddell seal, *Leptonychotes weddelli*, females are larger than males, yet males hold harems of up to twelve females, with a mean harem size of three, and fights between males for access to females are common (Mansfield, 1958; Kaufman, Siniff & Reichle, 1975; Jouventin & Cornet, 1979; Cornet & Jouventin, 1980). Similarly, females are larger than males in some polygynous bat species (Bradbury, 1977). It has been suggested that small size may enable males to be more mobile (Ghiselin, 1974; Kaufman *et al.*, 1975), and Ghiselin (1974) has suggested that small males might be able to spend more time and energy hunting for females.

In not all species will males be selected to maximize $(1 - t)W_m^c$. In these promiscuous species where size conveys no advantage in procuring access to females, males may be selected to maximize sperm production, and hence E_{rep}. This would lead to no sexual dimorphism in size. In some species, perhaps including some moths, males may be selected to maximize the distance they can travel. Suppose that male energy expenditure while trying to reproduce equals not $k_3 W^a(1 - t)$, but $k_7 W^a(1 - t) + k_8 W^e s(1 - t)$, where s = speed, $(1 - t)$ = time spent locomoting, k_7 and k_8 are constants and $k_8 W^e$ is the cost of locomotion per unit time per unit speed. Now $s(1 - t) = D$, distance travelled, and $e \simeq a$ (Fedak & Seeherman, 1979). Therefore we have

$$(k_2 W^b - k_1 W^a)t \simeq k_7 W^a(1 - t) + k_8 W^a D \qquad (7.24)$$

and so

$$D \simeq (k_2 W^{b-a} + (k_7 - k_1)t - k_7)/k_8 \qquad (7.25)$$

It is obvious from Equation (7.25) that D has its maximum at $W_m = 0$.

Sexual dimorphism and the energy available for reproduction

The ratio of E_{repm} to E_{repf}, the energy available to a male for reproduction divided by the energy available to a female for reproduction, depends on W_m/W_f, we can predict, as follows:

$$\frac{E_{repm}}{E_{repf}} = \frac{k_2 W_m^b - k_1 W_m^a}{k_2 W_f^b - k_1 W_f^a} \tag{7.26}$$

Substituting for k_2 from Equation (7.4), this reduces to

$$\frac{E_{repm}}{E_{repf}} = \left[\frac{a}{b}\left(\frac{W_m}{W_f}\right)^b - \left(\frac{W_m}{W_f}\right)^a\right] \Big/ \left(\frac{a}{b} - 1\right) \tag{7.27}$$

For $a = \frac{3}{4}$, $b = \frac{2}{3}$, the shape of this curve is shown in Figure 7.3.

The proportion of time a male spends trying to reproduce, $(1 - t)$, can be written as a function of W_m/W_f and K. Equating Equations (7.7) and (7.27) and substituting for k_2 from Equation (7.4), we have

$$1 - t = \left[\frac{a}{b} - \left(\frac{W_m}{W_f}\right)^{a-b}\right] \Big/ \left[\frac{a}{b} + (K - 1)\left(\frac{W_m}{W_f}\right)^{a-b}\right] \tag{7.28}$$

Values of $(1 - t)$ from Equation (7.28) for some values of W_m/W_f and K, for $a = \frac{3}{4}$, $b = \frac{2}{3}$, are listed in Table 7.6. When $(1 - t)$ is replaced by $(d - t)$, Equation (7.28), giving the proportion of time a male can spend trying to reproduce, becomes, on the assumption that $K = 1$:

$$d - t = d(1 - b(W_m/W_f)^{a-b}/a) \tag{7.29}$$

Table 7.6. *Predicted values of the proportion of time that males can spend trying to reproduce, $(1 - t)$, as functions of W_m/W_f, the degree of sexual dimorphism in body weight, and K, the ratio of k_3 to k_1*

		K			
		0.5	1.0	2.0	4.0
	0.25	0.344	0.208	0.116	0.062
	0.5	0.277	0.161	0.088	0.046
W_m/W_f	1.0	0.200	0.111	0.059	0.030
	2.0	0.110	0.058	0.030	0.015
	4.0	0.005	0.002	0.001	0.001

Note: These values of the proportion of time that males can spend trying to reproduce are calculated from Equation (7.28), assuming a $= \frac{3}{4}$, $b = \frac{2}{3}$.

so that, by comparison with Equation (7.28), the proportion of time a male spends trying to reproduce is now multiplied by a factor d. It is difficult to know what value d has. It will be greater than the proportion of time spent feeding and reproducing, as d includes both ingestion and digestion. A reasonable lower limit for d might perhaps be one-third.

Discussion

Equation (7.27) – see Figure 7.3 – predicts that in species where males are heavier than females, larger males should have less energy for reproduction. Searcy (1979) found that larger red-winged blackbird males perform less territorial behaviour. He claimed that this confirmed his hypothesis that larger males have less energy available for reproduction. However, larger males might perform less territorial behaviour even with larger energy stores than smaller males, because, being larger, each temporal unit of territorial behaviour is energetically more costly. There is some evidence that in certain dragonflies and damselflies, larger males may have lower reproductive success than smaller males. In the damselflies *Enallagma hageni* (Fincke, 1982) and *Coenagrion puella* (Banks & Thompson, 1985), intermediate-sized males were found to achieve a higher mating success than large or small individuals. Convey (1987) found that in the dragonfly *Libellula quadrimaculata* the larger satellite males had a lower mating success than the smaller territorial males. Convey (in prep.) suggests that larger males may be under 'some form of energetic constraint, such as relatively lower energy reserves'.

The model used in this chapter also predicts that above a certain weight, larger size in females should also be associated with a decrease in reproductive success. The only data of which I know that relate to this are those of Banks & Thompson (1987) for females of the damselfly *Coenagrion puella*. In a detailed long-term study of lifetime reproductive success, as measured by lifetime egg production, Banks & Thompson found that:

$$\text{eggs per clutch} = 699.19 - 6.4851 \text{ head width}$$

($r = -0.422$, $n = 25$) and:

number of clutches = 0.6137 head width − 45.45

(r = 0.219, n = 174). Multiplying these two equations together, an equation for the relationship between lifetime egg production and body size was obtained:

lifetime egg production = 723.81 head width − 3.98 head width^2 − 31778.6

This predicts that above a head width of 4.43 mm larger females have lower lifetime reproductive success. In reality female head widths varied between 3.65 and 4.4 mm.

Equations (7.28) and (7.29) predict that for most species ($K \simeq 1$; $0.5 \leqslant W_m/W_f \leqslant 2$; $d \simeq 0.5$) the proportion of time a male spends trying to reproduce should be about 0.03–0.1. Although some species show explosive breeding and breed for only a day or two each year (Wells, 1977, 1979), this seems very approximately to be

Figure 7.3. Predicted ratio of the energy available to a male for reproduction divided by the energy available to a female for reproduction as a function of sexual dimorphism in body weight; from Equation (7.27) for $a = \frac{3}{4}$, $b = \frac{2}{3}$.

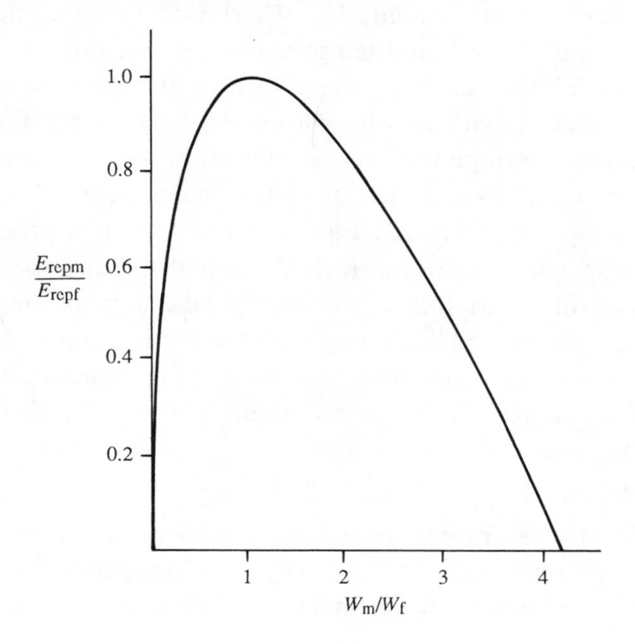

true. In grey seals, individual males spend up to 30 days trying to reproduce with a mean of 13 days (Boness & James, 1979). In the Antarctic Fur seal, *Arctocephalus gazella*, the mean period of tenure for bulls seen to copulate was 34.3 days (McCann, 1980). Figures of about 30 days are typical for many other seals (Gentry, 1973, 1975; Anderson, pers. comm.). Individual male European common toads spend up to about 4 weeks trying to reproduce, with a mean of 13 days per year at the breeding pool (Davies & Halliday, 1979). In Woodhouse's toad the corresponding figures are 25 and 8.5 nights (Sullivan, 1982), and in natterjacks they are about 35 days and 15 days (Arak, pers. comm.). In red deer, individual males spend up to about 4 weeks at the annual rut, with a mean of about 10 days (Clutton-Brock & Albon, in prep.).

Conclusions

The dependence of male reproductive success on body weight is investigated and incorporated into a quantitative model of sexual dimorphism in body size. It is concluded that the ratio of male to female body weight, W_m/W_f, should intraspecifically lie between 0 and 11, and an attempt is made to identify the relative importance of the factors that determine its exact value. This prediction accords with the observation that in a variety of taxa the most extreme examples of sexual dimorphism are found when females are larger than males, not when males outweigh females.

An equation is presented, Equation (7.20), which predicts the degree of sexual dimorphism in body weight for any species as a function of three variables: *a*, the intraspecific slope of log average daily metabolic rate on body weight; *b*, the intraspecific slope of log energy assimilation on log body weight; *c*, the intraspecific slope for males of log reproductive success on log body weight. For $a = \frac{3}{4}$, $b = \frac{2}{3}$, this equation reduces to:

$$W_m/W_f = \{27c/(24c + 2)\}^{12} \qquad (7.30)$$

For eight species, rather unsuccessful predictions of W_m/W_f are made. A reason is suggested for why in some species females are larger than males even though increased male size is associated with increased reproductive success. Essentially this is because, up to a certain size, females as well as males benefit from increased size.

The dependence of the energy available to a male for reproduction on his body weight is considered, and it is suggested that males can afford to spend only about 3–10% of their time trying to reproduce.

8

Are larger species more dimorphic in body size?

Many authors, from Darwin (1871) onwards, have considered factors that might affect the degree of sexual dimorphism. Two generalizations are frequently made. First, that dimorphism increases with the degree of polygyny (e.g. Crook, 1962; Lack, 1968; Clutton-Brock & Harvey, 1976; Clutton-Brock *et al.*, 1977; Alexander *et al.*, 1979) and, secondly, that larger species are more dimorphic (e.g. Rensch, 1950; Wiley, 1974; Ralls, 1976a; Leutenegger, 1978; Harvey & Mace, 1982). Intrasexual selection acting on males is usually given as the reason why males are larger than females in polygynous species (Selander, 1972; Belovsky, 1978). With increased polygyny increased sexual selection is expected, with the predicted consequence that the sexes should differ more in their optimal sizes. More recently, however, the role of selection for optimal female size has been emphasized, and advantages for smaller females have been postulated in polygynous species (Willner & Martin, 1985; Robinson, 1986).

The reasons, however, why larger species should exhibit greater sexual dimorphism in body size are less apparent than the reasons why more polygynous species should be more dimorphic. This chapter first reviews the evidence for the association between size and sexual dimorphism, and then considers those theories that predict that larger species should show greater dimorphism in body size.

A review of the evidence

Rensch (1950) and Rensch (1959) are frequently cited as evidence that sexual dimorphism is greater in larger species, but the data are unconvincing. No data are presented in Rensch (1959). In his 1950 paper, Rensch presented information from birds, mammals and carabid beetles. Pairs (occasionally triplets or quadruplets) of species were arranged by taxonomic affinity and from the data in his Tables 1–4 one can see whether the larger species in each comparison is more dimorphic. Sample sizes for each sex ranged from 5 to 47 in the beetles, and from 2 to 14 in the mammals, but were not recorded for the birds. No statistical tests were performed.

Rensch's data are summarized in Table 8.1. Before recording whether larger species are more or less dimorphic than smaller species, I have classified species according to which sex is the larger. This is because Rensch (1953) maintained that where females are larger than males, smaller species are more dimorphic in size, and because such 'reversed' sexual dimorphism in birds has been considered a separate phenomenon (Earhart & Johnson, 1970; Reynolds, 1972; Selander, 1972; Mosher & Matray, 1974; Walter, 1979; Wheeler & Greenwood, 1983).

Crump (1974) obtained data on 51 species of Ecuadorian anurans. There was no correlation between female length and sexual dimorphism in length: $r = 0.036$, $p > 0.5$.

Wiley (1974) showed a close correlation between female weight and sexual dimorphism in body weight across 11 species of grouse in the Tetraonidae (his Fig. 1): $r_s = 0.927$, $p < 0.001$.

Shubin & Shubin (1975) noted that sexual dimorphism in the Mustelidae (weasels, otters, etc.) is greatest in the *smallest* species. (For statistical analyses of the Mustelidae see below – Moors (1980) and Ralls & Harvey (1985).)

Ralls (1976a) noted that in mammals most extremely dimorphic species are large, and most occur in orders in which the modal species size is large, namely Primates, Pinnipedia, Proboscidea and Artiodactyla, while in the Marsupialia and the Chiroptera the extremely dimorphic species are found in the families that have the largest modal species size of their orders.

Table 8.1. *Presentation of data from Rensch (1950) on sexual dimorphism in size as a function of species size – data from pairs of species*

Taxon	Cases where males larger than females			Cases where females larger than males		
	Larger species more dimorphic	Smaller species more dimorphic	Species equally dimorphic	Larger species more dimorphic	Smaller species more dimorphic	Species equally dimorphic
Carabid beetles	0	0	0	5	3	0
Birds	12	3	0	2	2	1
Mammals	1	1	1	1	3	0

Note: Taken from Reiss (1986a).

Clutton-Brock *et al.* (1977) examined the dependence of sexual dimorphism on socionomic sex ratio and body weight in 42 primate species. Socionomic sex ratio, a measure of male–male competition for access to females, equals the mean number of adult females per adult male in breeding groups. Sexual dimorphism was positively related to body weight (the average of male and female body weight): $r_s = 0.484$, $p < 0.01$. This relationship was not a product of the inclusion of female body weight in both parameters – cf. Atchley (1978) – because when male weight was regressed on female weight, by least squares, the slope was significantly greater than one: $b = 1.063$, $p < 0.001$. Nor was this result a product of an association between large size and terrestriality since the same relationship held within arboreal species: $r_s = 0.395$, $p < 0.05$. Importantly, Clutton-Brock *et al.* also tested whether this result was caused by the known association between size and polygyny. Partial regression showed that the degree of sexual dimorphism in body weight was significantly related both to socionomic sex ratio ($b_1 = 0.111$, $p < 0.05$) and to body weight ($b_2 = 0.078$, $p < 0.005$). Indeed the dependence on body weight was considerably stronger than on socionomic sex ratio: standardized partial regression coefficients for socionomic sex ratio and body weight were 0.28 and 0.41 respectively. Similar results were obtained when the same analysis was applied to the data for arboreal primates only ($b_1 = 0.097, p < 0.1$; $b_2 = 0.064, p < 0.05$). When monogamous species were excluded from the analysis, no relationship was found between sexual dimorphism and socionomic sex ratio ($r_s = 0.087$, $t = 0.44$, ns).

Alexander *et al.* (1979) found strong correlations between sexual dimorphism and mean or maximal harem size in pinnipeds (15 species), ungulates (17 species) and primates (22 species). They wrote 'We tested for the possibility that increases in sexual dimorphism are partly a function of increasing body size and found no relationship'.

Mace (1979) tested whether sexual dimorphism increases with body size in small mammals. She concluded that in the Sciuridae, the Cricetidae, the Muridae and in her overall samples it did. This conclusion was reached by a least squares regression of female body weight on male body weight. The departure of this line from a predicted slope of one was tested by a *t*-test, because if larger species are more dimorphic such a slope should have a slope of less than one. (To normalize the data logarithms were used.) However, least squares regressions will have underestimated the true slopes of the regressions, as male and female body weight were presumably measured with approximately equal error, whereas least squares regression assumes that only the *y*-dependent variable is measured with error (Chapter 1). Georgina Mace kindly sent me the data to re-analyse, and a re-analysis is presented in Table 8.2, where the results of a reduced major axis regression of log W_m on log W_f are given. Although the overall departure of the slope from one is now not significant, the same three families still have slopes that differ significantly from one, and, in all, seven of the eight families have slopes greater than one.

Berry & Shine (1980) reported that for 11 species of kinosternid turtles (*Sternotherus* and *Kinosternon*) the ratio of male to female carapace length increases with the average of male and female carapace length: $r = 0.84$, $p < 0.005$.

Moors (1980) reported that in 24 samples from 15 mustelid species there was an inverse correlation between male body weight

Table 8.2. *Intergeneric dependence of male body weight on female body weight in small mammals by reduced major axis regression*

Family	Number of genera	Correlation coefficient	Slope	*t*
Sciuridae	10	0.999	1.072	4.193***
Geomyidae	3	0.995	1.032	0.321
Heteromyidae	5	0.992	1.123	1.501
Cricetidae	21	0.997	1.042	2.343*
Muridae	19	0.998	1.046	2.720**
Soricidae	7	0.960	1.159	1.098
Leporidae	3	0.998	0.949	0.810
All genera	88	0.997	1.012	1.331

Logged generic values are used. Slopes are tested for significance of departure from one. * $= 0.05 > p > 0.02$; ** $= 0.02 > p > 0.01$; *** $= 0.01 > p > 0.001$. Data from Mace (pers. comm.). Taken from Reiss (1986a).

and the degree of sexual dimorphism in body weight (Kendall rank test: $r = -0.35$, $p < 0.02$).

In a more complete analysis of mustelids, Ralls & Harvey (1985) used data from 26 species of family Mustelidae, belonging to 14 genera, to obtain a parametric correlation of sexual dimorphism in body weight and log male body weight using generic values (the means of values for species within the genus). Larger genera were again found to be less dimorphic ($r = -0.581$, $p < 0.05$), although this was not found to be the case when generic values of length rather than weight were used ($r = 0.281$, ns), nor was there any relationship within the genus *Mustela* (8 species) between sexual dimorphism in weight and log male weight ($r = -0.199$, ns) or between sexual dimorphism in length and log male length ($r = 0.165$, ns).

Stamps (1983) found that in an analysis of territorial lizards there was no indication of a relationship between female snout-vent length and sexual dimorphism in snout-vent length ($r = 0.066$, 29 df).

Arak (pers. comm.) found that for 18 species of American toads in the genus *Bufo* there was no correlation between male length and the ratio of male to female length: $r = -0.007$, $p > 0.5$.

Why might larger species be more dimorphic?

In some taxa the interspecific relationship between size and the degree of male–male competition may explain the interspecific relationship between size and the degree of sexual dimorphism. For example, Wiley (1974) showed that larger species of grouse were not only more dimorphic, but also more polygynous. Larger species are also more polygynous in antelopes (Jarman, 1974), primates (Clutton-Brock *et al.*, 1977) and deer (Clutton-Brock, Albon & Harvey, 1980). In ungulates and primates food distribution is probably responsible for small species being less polygynous (Jarman, 1974; Clutton-Brock & Harvey, 1977). In mustelids the greater size dimorphism in the smaller species (Shubin & Shubin, 1975; Moors, 1980; Ralls & Harvey, 1985) may be caused by the fact that the smaller species are probably more polygynous (see Lockie, 1966; Corbet & Southern, 1977).

The most convincing data to show that larger species are more dimorphic, even after the removal of any interspecific correlation between size and inter-male conflict, come from primates (Clutton-Brock *et al.*, 1977) and small mammals (Mace, 1979; Table 8.2). The small mammal data are the less convincing as arboreal genera tend to be small and also less dimorphic (Mace, 1979). When Mace repeated the analyses within zonation classes the significant relationships were lost, although this could have been because of the small sample sizes involved. Mace discounted the possibility that larger genera are also more dimorphic because polygynous species tend to be both larger and more dimorphic on the grounds that her analysis did not show that polygynous species are larger. As she admitted, however, information on which to make such social classifications for small mammals is scanty.

Several theories have been put forward to predict a positive correlation between dimorphism and size.

Rensch (1959) invoked allometry: 'Secondary sex characters of many mammals usually represent structures of positively allometric growth because they usually become more conspicuous during a more advanced period of ontogenetic development (due in mammals to the increasing influence of sex hormones).' However, as Clutton-Brock *et al.* (1977) point out, there is no obvious reason to suppose that allometry is inevitable (cf. Clutton-Brock & Harvey, 1979; Clutton-Brock *et al.*, 1980).

Wiley (1974) suggested that because the evolutionary advantages of deferred reproduction – and thus larger size at maturity – probably increase in saturated environments (MacArthur & Wilson, 1967; Pianka, 1972; MacArthur, 1972) sexual dimorphism in size, and the tendency in grouse for larger species to be more dimorphic, would evolve if large size and delayed reproduction did not have the advantages for females that they do for males. Wiley suggested that this would be so 'If the amount of food a larger female grouse could collect each day does not increase in proportion to the increase in her existence energy requirements'. This seems unconvincing as the same argument applies to males.

Maynard Smith (1977) suggested that sexual selection for increased size in males may, as a side effect, have led to increased female size, thus creating a relationship between dimorphism and average body size, Clutton-Brock *et al.* (1977) raised two objections. First, that Maynard Smith's explanation requires that increases in male size are more closely paralleled by increases in female size in small animals than in large. Secondly, that selection for energetic efficiency probably imposes strict constraints on female size.

Clutton-Brock *et al.* (1977) suggest that larger species have fewer competing species, thus relaxing constraints on increasing male size imposed by the presence of larger species in neighbouring niches, so allowing sexual selection for male competitive ability to produce relatively larger increases in male size. In fact the evidence that body size is constrained by interspecific competition is contentious (Dunham, Tinkle & Gibbons, 1978; Wiens & Rotenberry, 1981; Simberloff & Boecklen, 1981). However, if this is the case, we can envisage the following evolutionary scenario.

Species are arranged along a one-dimensional resource axis with body weights W_o, xW_o, x^2W_o, x^3W_o, . . . , where $x \simeq 2$ (Hutchinson, 1959; Brown, 1975; Diamond, 1975; Uetz, 1977). The nth species, of weight $x^{n-1}W_o$, now becomes sexually dimorphic and the degree of sexual dimorphism is constrained by the weights of the neighbouring species. If the neighbouring species remain at weights $x^{n-2}W_o$ and x^nW_o, then, assuming males are heavier than females, we have

$$x^{n-2}W_o/W_f = W_f/W_m = W_m/x^nW_o$$

where W_f is adult female body weight and W_m is adult male body weight. Therefore

$$x^{n-2}W_oW_m = W_f^2$$

and

$$x^nW_oW_f = W_m^2$$

i.e.

$$W_m/W_f = x^{\frac{2}{3}} \simeq 1.59 \quad \text{for} \quad x = 2.$$

So W_m/W_f is predicted to be independent of species size. If we assume $W_f > W_m$, then $W_f/W_m = x^{\frac{2}{3}}$. If all the species on the resource axis become sexually dimorphic then W_m/W_f or $W_f/W_m = x^{\frac{1}{2}} \simeq 1.41$ for $x = 2$. Again the degree of sexual dimorphism is predicted to be independent of species size. For the 12 North American falciform taxa listed by Selander (1972) sexual dimorphism in wing length is indeed independent of species size ($r_s = -0.238$, $p > 0.2$) and W_m/W_f – assuming body weight proportional to (wing length)^3 – ranges from 1.08 to 1.66 with a mean of 1.35 not significantly different from 1.41 ($t = 0.401$, $p > 0.5$). It is still not certain why in predatory birds the sexes differ in size, nor why in such species females are always larger than males (Newton, 1979; Walter, 1979; Wheeler & Greenwood, 1983), but the above analysis suggests that we cannot rule out the possibility that in such species the evolution of sexual dimorphism is constrained by interspecific factors. If x increased with species size, then the theory of Clutton-Brock *et al.* would be tenable. Oksanen, Fretwell & Järvinen (1979) indeed argued that in some, but not all, avian communities there is a gap of two or three species between the largest and next largest species. This theory has, however, been convincingly criticized (Wiens & Rotenberry, 1981) and the phenomenon does not seem to occur in other taxa: spiders (Uetz, 1977), rodents (Brown, 1975), birds (Diamond, 1975).

Clutton-Brock *et al.* (1977) suggest that 'energetic constraints on increasing male body size may be relaxed in larger species since larger body size leads to a decrease in the surface area/volume ratio, metabolic rate and relative nutrient requirements . . . thus making further increases energetically "cheaper"'. This theory is contradicted by the analyses of Chapter 7. Figure 8.1 plots the intraspecific curves for non-reproductive energy requirements, $k_1 W^a$, and energy assimilation, $k_2 W^b$, for individuals of various body weights, on the assumption that $a = \frac{3}{4}$, $b = \frac{2}{3}$. For the purpose of this chapter, the relevant point is that the inequality for W_m/W_f predicts no dependence of W_m/W_f on species size:

$$0 \leqslant W_m/W_f \leqslant (a/b)^{1/(a-b)} \tag{7.19}$$

Pace Clutton-Brock *et al.* (1977) such energetic arguments do not predict that in larger species further increases in size are energetically 'cheaper'.

Clutton-Brock *et al.* (1977) also suggest that 'selection may

Figure 8.1. Non-reproductive energy requirements, $k_1 W^a$, and energy assimilation, $k_2 W^b$, for individuals of different body weights. On the assumption that these curves hold intraspecifically both for males and for females, the degree of sexual dimorphism in body weight, W_m/W_f, is predicted to be independent of species size and to lie between 0 and about 4, if $a = \frac{3}{4}$ and $b = \frac{2}{3}$.

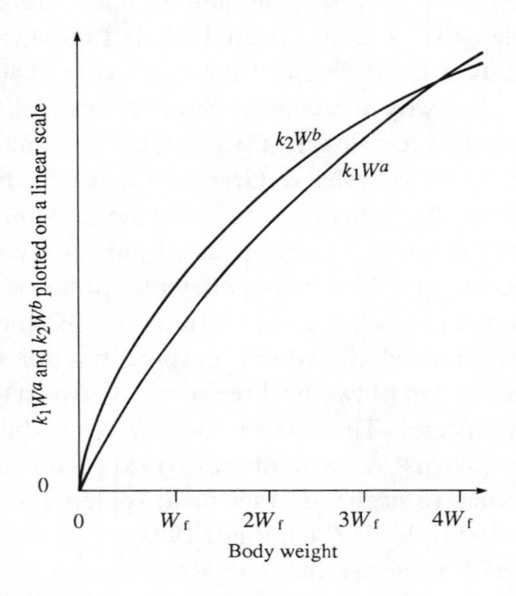

favour increased feeding divergence (and thus increased dimorphism) in species which feed on low density food distributed in clumps, since, in this situation, the effects of feeding competition are strongest. Mammalian species which feed on dispersed and unpredictable food supplies are often of large size'. However, selection for feeding divergence between the sexes is likely to be most important in monogamous species: in such species there is no evidence that larger species are more dimorphic (e.g. Fig. 6 of Clutton-Brock & Harvey (1977) for primates.)

It has been argued that within a species smaller females may be able to breed at a younger age than larger ones (Downhower, 1976). Clutton-Brock *et al.* (1977) suggest that this pressure might be more acute in large species where the age at first breeding is relatively late, than in small species where it is generally early. It could, however, equally be argued that the pressure for breeding early is more acute in small species with their *r*-selected characteristics (MacArthur & Wilson, 1967; Pianka, 1970; Western, 1979).

Moors (1980) suggested that in female mustelids small size favours reproductive efficiency whereas large size favours increased survival; 'These consequences of optimal reproductive strategy have the effect of favouring prominent sexual dimorphism in species with relatively brief lives. Usually such mustelids are also small. Thus reproductive strategies may be one of the factors causing the inverse relationship between body size and dimorphism'. It does not seem apparent why small size should favour 'reproductive efficiency' (cf. Figure 8.1). As suggested earlier, one reason why larger mustelids are less dimorphic may be that they are less polygynous.

Discussion

In several taxa, larger species are more dimorphic. In only primates (Clutton-Brock *et al.*, 1977) and rodents (Mace, 1979; Table 8.2) is there evidence that this effect exists even when the tendency for larger species to be more polygynous, and hence more dimorphic, is removed, and it could be that these two studies do not manage entirely to remove the effect of size on polygyny. As Mace (1979) noted, the information on rodent social organization is poor, and perhaps in primates socionomic sex ratio is an inadequate indicator of sexual selection.

As Clutton-Brock *et al.* (1977) point out, none of the explanations for why larger species should be more dimorphic is wholly satisfactory. Indeed all of the proposed explanations can be criticized. Other theories will no doubt be proposed. It could be that as larger species have reserves that enable them to survive for longer without feeding (Brodie, 1975; Kleiber, 1975), males in larger species can afford to be nearer the upper limit of body size, where $k_1 W^a$ equals $k_2 W^b$, and so larger species are more dimorphic in size. As discussed in Chapter 5, in some taxa larger species live in larger social groups, and therefore may be predisposed to be more polygynous.

Even in those taxonomic groups where there is statistical

evidence that larger species are more dimorphic, one of the perennial problems of correlation tests remains; namely that the data used within any one correlation are not necessarily independent as the statistical test requires (Sokal & Rohlf, 1969). This point is especially pertinent for those correlation studies carried out by using species (rather than genera) as their raw data (Harvey & Mace, 1982; Ridley, 1983).

Conclusions

It is frequently stated that larger species show greater sexual dimorphism in size. Here the evidence for this statement is reviewed and the theories that predict an association between size and dimorphism considered. In several taxa larger species are more dimorphic, although there are many exceptions. Indeed in the Mustelidae (weasels, otters, etc.) smaller species are more dimorphic.

The main reason why an association between size and sexual dimorphism sometimes exists is probably because, on an evolutionary timescale, ecological factors such as food distribution affect both size and the opportunity for polygyny – polygynous species tend to be dimorphic – rather than because of a direct causal link between size and dimorphism. When the effects of polygyny on sexual dimorphism are removed, in only primates and small mammals is there still evidence of a link between size and dimorphism.

The theories that predict a causal link between size and dimorphism are generally unconvincing.

9

Surface area/volume arguments in biology

Surface area/volume arguments have been extensively used in biology. They are valid provided, first, that one of the properties being considered scales as the surface area of an organism and the other as the volume of an organism and, secondly, given that this is the case, that the surface area/volume ratio represents an important evolutionary constraint pertinent to the problem under consideration.

This chapter looks at some instances where surface area/volume arguments may have been misappropriated or misapplied. It also considers when they may be valid.

Why cannot large animals rely solely on diffusion for gaseous exchange?

The universal answer given in school (Mackean, 1973; Revised Nuffield Biology, 1975; Soper & Tyrell Smith, 1979; Rowlinson & Jenkins, 1982; Jones & Jones, 1984; Green, Stout & Taylor, 1984; Roberts, 1986a, b; Hill & Holman, 1986) and university (Marshall & Hughes, 1965; Barrington, 1967; Alexander, 1971, 1979; Barnes, 1974; Bligh, Cloudsley-Thompson & MacDonald, 1976; Schmidt-Nielsen, 1983, 1984) text-books to the question 'Why do large species of animals have lungs, gills or other structures specialized for gaseous exchange while smaller species manage by diffusion' is that the larger an organism, the smaller its surface area to volume ratio. This answer makes two implicit assumptions.

First, that the oxygen requirements of an animal are proportional to its volume or weight. Secondly, that diffusion can only provide oxygen at a rate proportional to an animal's surface area. The former of these assumptions is incorrect (Reiss, 1987b). The oxygen requirements of animals scale on body weight with exponents of less than one. Metabolic requirements are proportional to $W^{0.5-0.8}$, not $W^{1.0}$ (Chapter 2). Consequently the surface area/volume argument does not explain why only small animals rely on diffusion alone to supply their oxygen needs.

Had metabolic rates scaled as $W^{1.0}$, then an eightfold increase in body weight would lead to a doubling in the amount of oxygen required per unit area of body surface per unit time on the assumption that surface areas scale as the two-thirds power of body weight. If, however, oxygen requirements scale as basal metabolism perhaps does, as $W^{0.75}$, then an increase in body weight of a factor of 4000 would not even double the amount of oxygen required per unit area of body surface per unit time!

In the real world, average daily metabolic rates, ADMR, or even maximal rates of oxygen consumption, may better reflect the constraints against which respiratory structures have evolved than do basal metabolic rates. As noted in Chapter 2, the exponents relating these measures have, perhaps unsurprisingly, been determined for fewer groups of organisms and with less precision. The most careful analyses show that in birds and mammals, for which most data exist, each of these measures scales on body weights with exponents ranging from 0.5 to 0.8 (Prothero, 1979; Taylor *et al.*, 1980; Peters, 1983; Chapter 2). These conclusions hold whether least squares or reduced major axes regressions are used (Reiss, 1982a).

The surface areas of organisms scale on body weight with exponents close to the predicted value of two-thirds. Measured exponents range from 0.65 to 0.74 (Peters, 1983).

It is important to realise that the above argument makes the implicit assumption that diffusion occurs through an epithelium, whose thickness is either independent of body size or so small as not significantly to impede the passage of respiratory gases, whereupon a circulatory system carries the oxygen round the body. The data most pertinent to this assumption came from Gehr, Mwangi,

Ammann, Maloiy, Taylor & Weibel (1981). In a study of 15 African mammals spanning a range in body mass from 0.4 to 250 kg they found that mean barrier thickness between air in the alveoli and blood in capillaries ranged only from 0.37 to 0.65 μm and scaled on body weight with an exponent of 0.03, which was not significantly different from 0. When the number of species was increased to 33, with a consequent range in body mass from 0.002 to 700 kg, by the additional use of data obtained from the literature, mean barrier thickness still ranged only from 0.26 to 0.65 μm, scaling on body weight with an exponent of just 0.05, although this was now significantly different from 0.

Given that the oxygen requirements of organisms do not follow a volume law, why is it that large animals have specialized respiratory structures? At least four possible explanations can be suggested.

First, even if oxygen requirements do not scale as $W^{1.0}$, it is still possible that the exponent relating oxygen requirements to body weight is greater than the exponent relating surface areas to body weight.

Secondly, as noted by Hemmingsen (1960), a multicellular poikilotherm has a basal metabolic rate almost 10 times that of a unicellular organism of the same size, and a homeotherm's basal metabolic rate is almost thirty times that of an equivalent sized poikilotherm. These 'quantum leaps' in oxygen demands evidently have nothing to do with a surface area/volume law, but do show that it appears as if diffusion alone is totally inadequate to supply the oxygen requirements of birds and mammals. This argument loses some of its force when it is realized that maximal rates of aerobic energy expenditure in poikilotherms may be up to 300 times basal metabolic rate, whereas in homeotherms maximal rates of oxygen consumption appear only to exceed basal rates by factors of about 10 (Hemmingsen, 1960; Taylor *et al.*, 1980; Peters, 1983). As first pointed out by Hemmingsen, this means that the maximal metabolic rates of insects in flight are about the same as those that are expected from similarly sized homeotherms exercising at their maximal rates. Of course, neither the jump from unicellular organisms to poikilotherms, nor that from poikilotherms to homeotherms, corresponds to a change from surface diffusion to respiratory organs. The salamanders are a taxon that contains both lunged and lungless forms. Whitford & Hutchinson (1967) studied four species of lungless salamanders, ranging in mass from 3.5 to 22 g, and seven species of lunged salamanders, ranging in mass from 3 to 31 g. At the same temperature the lunged salamanders had significantly higher metabolic rates than the lungless ones by between 10 and 50%.

Thirdly, large animals obviously have thick skins for protection. Consequently much of their body surface area is unsuitable for gaseous exchange by diffusion.

Finally, large animals simply have more room for lungs, gills or other specialized respiratory structures.

Discussion

The above argument does not apply to organisms such as flatworms, which lack a circulatory system and in which diffusion has consequently to take the oxygen all the way to the tissues. For such organisms, we can show, following the methodology of Alexander (1971, 1979), that the requirement for diffusion to be adequate to supply the oxygen needs of an organism, whatever its size, is for the metabolic needs of that organism per unit time to scale on its body weight with an exponent of 0.33 or less.

Imagine a spherical organism of radius r. Consider that part of the organism that lies between a distance x and $x + \delta x$ from the centre (Figure 9.1). We have

$$\begin{aligned}\text{volume between } x \text{ and } x + \delta x &= 4\pi\{(x + \delta x)^3 - x^3\}/3 \\ &\simeq 4\pi x^2 \delta x\end{aligned}$$

Suppose oxygen requirements of the whole organism are $k_1 W^a$ per unit time. Then the oxygen requirements of the organism per unit volume are proportional to $k_1 W^{a-1}$. Consequently the oxygen requirements per unit time of the volume between x and $x + \delta x$ are proportional to

$$4\pi x^2 \delta x \cdot k_1 W^{a-1} \qquad (9.1)$$

Oxygen will be supplied at a rate (in units of volume per unit time) given by

$$J = -AK\mathrm{d}p/\mathrm{d}s \qquad (9.2)$$

where J is the rate of diffusion, A is the area of the surface across which the diffusion is occurring, K is a quantity known as the permeability constant, p is the partial pressure and s is the distance over which the oxygen is having to diffuse.

In our case we have $A = 4\pi x^2$ and $s = r - x$. Substituting these values for A and s into Equation (9.2) and equating Equations (9.1) and (9.2), we arrive at

$$\mathrm{d}p/\mathrm{d}s \propto k_1 W^{a-1} x/3K \qquad (9.3)$$

Writing p_e for the partial pressure of oxygen in the environment in which the animal is living and p_o for the partial pressure of oxygen at the centre of the animal, we have

$$p_e - p_o \propto \frac{k_1 W^{a-1}}{3K} \int_0^r x\delta x$$

Figure 9.1. A spherical organism of radius r, whose oxygen needs are supplied by diffusion. The oxygen requirements per unit time of the volume between x and $x + \delta x$ are proportional to Equation (9.1).

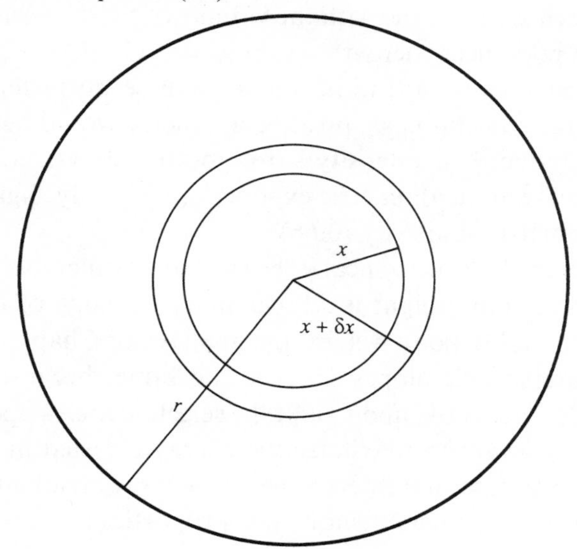

that is

$$p_e - p_o \propto k_1 W^{a-1} r^2/6K \qquad (9.4)$$

We can use the fact that r^2 will be proportional to $W^{\frac{2}{3}}$. Equation (9.4) now becomes

$$p_e - p_o \propto k_1 W^{a-\frac{1}{3}}/6K \qquad (9.5)$$

Because p_o cannot be negative, we have that the condition for diffusion to suffice for an organism with neither specialized respiratory structures nor a circulatory system is for

$$a < \tfrac{1}{3} \qquad (9.6)$$

Hypsodonty

Hypsodonty is the phenomenon that in many orders of mammals larger species have relatively taller teeth. The most frequent explanation for hypsodonty, complication of enamel ridge folding and expansion of the grinding surface area by more than would be predicted via isometry, e.g. by molarization of premolars or by positive allometry of molar surface grinding area, is that without these evolutionary strategems animals would not have enough tooth area to cope with the volume of food ingested (Gould (1966) and references therein). As, however, Gould (1975b) points out 'the demands of nutrition might increase no faster than the metabolic rate; in this case, positive allometry would be expected, but much reduced in intensity – for tooth area vs. body weight would increase at an allometric exponent (0.75) only slightly above that of geometric similarity (0.66)'.

In fact, even if intraspecifically the slope of log metabolic requirements on log body weight is steeper than the slope of log energy assimilation on log body weight, the analyses of Chapter 5 suggest that the interspecific slopes may be the same, because of interspecific differences in k_1 and k_2 and the selective forces operating on female body weight to maximize the energy invested in reproduction. If it is the case that interspecifically energy assimilation scales as a surface area phenomenon, i.e. proportional to $W^{0.67}$, why might larger species of mammals exhibit hypsodonty and have relatively larger and more complex teeth?

One possibility is that although energy assimilation per unit time may scale interspecifically as approximately $W^{0.67}$, larger species live for longer, so that over the course of an individual's lifetime, energy assimilation will scale on body weight with an exponent of greater than 0.67, possibly of 1.0. A second possibility is that larger species feed on poorer quality food, thereby wearing down their teeth relatively more quickly. A third possibility is simply that larger species have mouths large enough for relatively larger teeth.

Other surface area/volume arguments that may be invalid

It is a general phenomenon that larger organisms are structurally more complex, as discussed above with reference to respiratory surfaces and teeth. Such increasing complexity is often argued to be the consequence of some surface area/volume ratio, but may in fact simply be a consequence of larger organisms having a greater internal volume. Larger mammals show more complicated folding of the intestines (Thompson, 1942; Gould, 1966) and larger brachiopods show increased looping and folding of their food-gathering lophophores (Elliott, 1948). In each case it is difficult to construct a valid surface area/volume argument.

Bergmann's Rule states that 'races of warm blooded vertebrates from cooler climates tend to be larger than races of the same species from warmer climates' (Mayr, 1956). The classical evolutionary explanation of the rule is that larger animals have an advantage in cooler climates because their smaller surface area/volume ratios lead to lower rates of heat loss and metabolism per gram body weight (reviewed by Searcy, 1980). Objections have been raised to this explanation of Bergmann's rule. For one thing, variation in energy assimilation as well as energy requirements must be considered. In a careful review Geist (1987) concludes, in common with a number of other authors, that Bergmann's rule is invalid. In large mammals body size at first increases with latitude, but then reverses between 53° and 65°N, so that the largest individuals occur at these intermediate latitudes. Geist concludes that Bergmann's rule has no basis in fact or theory.

When are surface area/volume arguments valid?

If a careful check is made to ensure that the properties under consideration do scale either as the surface area or as the volume of an organism, some surface area/volume arguments can be seen to be valid.

Larger organisms take longer either to be warmed up by an external source of heat or for their body temperatures to fall by a stated number of degrees as a result of their losing heat to a cool environment, whether by convection, conduction or radiation. This is because heat gain or loss is proportional to the surface area of the organism, whereas the amount of heat that is gained or lost is proportional to the volume of the organism. This argument does not take into account the fact that larger organisms have thicker insulating layers. When these are taken into account, larger organisms will take even longer, relative to smaller ones, to warm up or cool down by a specific amount, although the relationship is now, strictly, not a surface area to volume one.

Methods of locomotion that rely exclusively on the surface of an organism, such as the beating of cilia, are less likely in larger organisms as the mass of the organism to be moved is proportional to its volume.

There are a number of surface area/volume ratio arguments that are strictly metabolic rate/volume ratio arguments. As, however, metabolic rates scale close to $W^{0.67}$ (Chapter 2), the two sorts of argument produce essentially the same predictions.

Consider, for example, the risk of starvation when no food is available. In this case, large size is an advantage. On the assumption that body reserves are proportional to body mass (Chossat's Law – Kleiber (1975)), and that metabolic rate scales as W^a, time to starvation will scale as W^{1-a}, so that larger animals survive for longer. In circumstances where the risk of starvation is caused not by a total cessation of food, but by a substantial reduction in its availability, surface area/volume arguments are less use. After all, smaller animals have the advantage of needing absolutely less food and may therefore be able to survive while larger animals starve. Similar considerations apply to the problem of dehydration.

Conclusions

Surface area/volume arguments are valid provided, first, that one of the properties being considered scales as the surface area of an organism and the other as the volume of an organism and, secondly, given that this is the case, that the surface area/volume ratio represents an important evolutionary constraint pertinent to the problem under consideration.

A surface area/volume argument is usually invoked as the explanation for why only small organisms can rely on diffusion for their supply of O_2 and removal of CO_2. This argument makes the implicit assumption that an animal's oxygen requirements are proportional to its volume. It is known, however, that this is not the case. Larger organisms require less oxygen relative to their body weights. The reasons why larger organisms have specialized respiratory surfaces has nothing to do with surface area/volume ratios.

For organisms that lack a circulatory system and in which diffusion has consequently to take the oxygen all the way to the tissues it can be shown that the requirement for diffusion to be adequate to supply the oxygen needs of an organism, whatever its size, is for the metabolic needs of that organism per unit time to scale on its body weight with an exponent of 0.33 or less.

Various hypotheses can be advanced for why larger species of mammals exhibit hypsodonty and have relatively larger and more complex teeth, but the phenomenon is not caused by surface area/volume effects. Similarly, Bergmann's rule and the tendency for larger organisms to have more complicated food-gathering structures and digestive surfaces are unlikely to have much to do with surface area/volume ratios.

Some surface area/volume arguments are valid, such as those to do with heat gain and loss and those to do with the effectiveness of surface structures such as cilia for locomotion. Others, for example the time taken to starve as a function of body size in the complete absence of food, are strictly consequences of metabolic rate/volume ratios and so approximate to surface area/volume arguments. The consequences of increased size on the likelihood of death by starvation or desiccation are not so straightforward, however, when there is a shortage rather than a complete absence of food or water.

10

Prospectus

This chapter mainly considers areas of major ignorance to which allometric arguments might make a valuable contribution. Possible shortcomings of the allometric approach used throughout this book are also voiced.

Allometry in plants

With a few notable exceptions, remarkably little work has been done on the allometry of plant growth and reproduction. Despite some early allometric approaches (Pearsall, 1927; Turrell, 1961) more recent papers and books on these topics (e.g. Hunt & Parsons, 1974; Hurd, 1977; Hunt, 1978; Venus & Causton, 1979) fail to include allometric analyses and instead abound with detailed calculations of plant growth curves, in particular; yet the plethora of equations and statistical techniques used fail to provide a functional framework within which to consider plant growth and reproduction.

A first approach would be to see whether the allometric growth equation of Chapter 5, Equation (5.4), provides a useful fit to plant growth curves.

A very great deal more could also be done on the allometry of plant reproduction along the lines of Whittaker & Woodwell (1968) and Hubbell (1980). Hubbell found that a doubling of circumference in the tropical tree *Bursera simaruba* produced nearly a

50-fold increase in mean seed crop (cf. Chapter 4). Such studies, whether intraspecific or interspecific, are easy and relatively quick to carry out, produce invaluable data and may be of commercial value too.

Optimal organ size

As one might expect, the weights of many organs scale both intraspecifically and interspecifically on body weight with exponents of close to one. There are, however, some interesting and notable exceptions.

Brain weights have long been known to scale on body weights with exponents of less than 1.0 (Jerison, 1973; Martin, 1981). Further, significant differences exist both between taxonomic categories and between ecological categories in encephalization quotients (Eisenberg & Wilson, 1978) or comparative brain sizes (Harvey, Clutton-Brock & Mace, 1980). For example, in both primates and small mammals, animals in the combined category of frugivores, insectivores and granivores have larger brains than would be expected for species of their body weights if they fed on leaves (Harvey *et al.*, 1980). The advantage of increased brain size is presumably that it allows the greater intelligence, coordination or memory that specialization on such foods may require. One imagines that increased brain size carries with it the disadvantage of greater metabolic requirements. Brains require huge amounts of glucose and oxygen for their size relative to all other organs except active skeletal muscle. It is perhaps within the bounds of possibility that one could usefully model the advantages and disadvantages of changes in brain size for an individual, thus predicting optimal changes in brain size.

Within the Cervidae (deer), larger species have relatively larger antlers even when the fact that larger species are more polygynous is taken into account (Clutton-Brock *et al.*, 1980). Conceivably within-species observations on natural variation in antler size and male reproductive success combined with energetic considerations of the annual cost of antler production might be able to identify the various selective forces operating on antler size. The high correlation between antler size and body weight (Huxley, 1932), however, is likely to complicate matters.

Across species analyses of mammals show that larger males have relatively smaller testes, although for species of approximately the same body weight, the males of even closely related species may differ considerably in testes size (Harcourt, Harvey, Larson & Short, 1981; Harvey & Bennett, 1985; Kenagy & Trombulak, 1986; Dixson, 1987). This variation is largely because of sperm competition. Similar considerations probably apply in many other taxa and to other organs, for example canines (Harvey, Kavanagh & Clutton-Brock, 1978a, b). Again, energetic and allometric studies and calculations, along the lines of those proposed above for optimal antler size, and carried out in Chapter 7 for optimal male body weight, may help determine the precise evolutionary pressures acting to maintain these organs at their particular sizes.

Following Wade & Arnold (1980), the definition of sexual selection on males as 'the variance in the number of mates per male' and on females as 'the variance in the number of mates per female' has become popular. Personally, I do not find this definition very useful as it tells us nothing about the outcome of such sexual selection. It does allow across-species rankings of the strength of sexual selection, as thus defined, to be made, but it does not tell one anything about why particular morphological and behavioural traits that differ between the sexes are maintained by natural selection at their observed levels.

Allometry of parental investment

Although we know an increasing amount about the energetics of growth and reproduction (e.g. Loudon & Racey, 1987), few studies make quantitative predictions about anything. Most of the predictions throughout this book operate at what might be termed the 'macroscale'. For instance, predictions for any group of animals about the scaling on size of the energy females invest in reproduction, of the duration of parental investment and of growth efficiencies. What will increasingly become feasible are predictions at the 'microscale', that is, predictions that apply only to one or perhaps a handful of species.

Altmann (1980), for example, provides an allometric analysis of maternal feeding time in yellow baboons, *Papio cynocephalus*. She concludes that a mother could not provide all the caloric requirements for herself and her infant beyond six to eight months of infant age and probably could do so up to that age only with difficulty and major restructuring of other aspects of her life.

The sorts of questions that could be addressed by similar allometric approaches include:

How is offspring weight at the end of parental investment affected by the rate and duration of parental investment?

What are the consequences to mother and offspring of changes in clutch or litter size?

What are the shapes for sons and daughters of the curves that relate offspring fitness to parental investment?

In species where parental investment continues after birth, what are the selective forces operating on the timing of birth?

What are the energetic consequences to both parent and offspring of offspring being carried around by a parent?

Some difficulties in the allometric approach

Theories can never be proved, only disproved, and it may be that some of the explanations presented here predict the right answer for the wrong reason. Any model makes simplifying assumptions. Models, such as those developed here, which claim considerable generality, are bound to prove oversimplified once applied to particular instances. In a sense, all the models that I present are specific instances of more general ones presented by Sibly & Calow (1986). The difficulty with very general models, such as Sibly & Calow's, is that their predictions may be so general as to be intuitively almost obvious. The difficulty with more specific ones, such as the energetic ones that I present, is that they leave out factors that may be of great importance. For instance, the predictions in Chapter 7 of the degree of sexual dimorphism in body size fail to consider such complications as:

Larger males may suffer more predation;

Larger males take longer to reach reproductive maturity;

If adult male size is determined to some extent by parental investment, parents may restrict their sons to sub-optimal (from the point of view of the sons) sizes, so as to enable themselves to invest in other offspring;

In seasonal environments, large males carry with them the disadvantage of smaller energy reserves with which to survive bad conditions;

Data obtained from within-population studies of the dependence of reproductive success on male size may not accurately reflect the options open to a particular male. Intraspecific variation within adult size might be at least partly adaptive.

Concluding discussion

'Ecology is still a branch of science in which it is usually better to rely on the judgement of an experienced practitioner than on the predictions of a theorist. Theory has never played the role that it has in population genetics, perhaps because there is nothing in ecology comparable to Mendel's laws in genetics' (Maynard Smith, 1974).

Allometry is a powerful tool, and although it would perhaps be an exaggeration to claim that its role in evolution, ecology, behaviour and physiology will ever be as central as the role of Mendel's laws in genetics, the basic equations of allometry nevertheless share with Mendel's laws features of simplicity and generality. There is, I suspect, much that allometry has still to contribute to biology.

Glossary of mathematical terms

α — Constant. See Equation (1.1).

a — Exponent relating non-reproductive energy requirements to body weight. Also used, in Chapters 1 and 4, as a constant.

A — Constant. Also used, in Chapter 5, to represent assimilation and, in Chapter 9, to represent the area across which diffusion is occurring.

ADMR — Average daily metabolic rate.

β — Exponent. See Equation (1.1).

b — Exponent relating energy assimilation to body weight. Also used, in Chapters 1 and 8, as a more general exponent relating some function of body weight to body weight, in Chapter 4 as an exponent relating fecundity to body weight and, in Chapter 6, as a constant.

b_1 — Partial regression statistic. See Chapter 8.

b_2 — Partial regression statistic. See Chapter 8.

B — Constant. Also used, in Chapter 5, to represent biomass and, in Chapter 6, to represent food bulkiness (= wet weight/dry weight).

BMR — Basal metabolic rate.

c — Constant. Also used, in Chapter 5, to represent k_1/k_2 and, in Chapter 7, to represent the slope of the regression of log reproductive success on log male body weight.

c' — c/r in Chapter 7.

c_1 — Constant.

c_2 Constant.
c_3 Constant.
c_4 Constant.
c_5 Constant.
C Thermal conductance. Also used, in Chapter 5, as a constant that relates energy stored to weight gain.
δ Constant.
δt Small unit of time.
δW Small unit of weight.
δx Small unit of distance.
d Diameter.
$d - t$ The proportion of the time males spend trying to reproduce.
df Degrees of freedom.
D Population density. Also used, in Chapter 6, to represent percentage digestibility and, in Chapter 7, to represent distance travelled.
DEB Daily energy budget.
DEE Daily energy expenditure.
e Exponent relating cost of locomotion per unit time per unit speed to body weight.
ECP Energy cost of play.
E_{in} Energy assimilation per unit time.
E_{gro} Energy available for growth per unit time.
$\bar{E}_{gro}/\bar{E}_{in}$ Proportion of the total energy budget devoted to growth from birth to adulthood.
$E_{gro+rep}$ Energy available for growth and reproduction per unit time.
E_{rep} Energy available for reproduction per unit time.
E^*_{rep} Greatest intraspecific value of E_{rep}.
E_{repf} Energy available to a female for reproduction per unit time.
E_{repm} Energy available to a male for reproduction per unit time.
E_{req} Energy requirements per unit time for everything except reproduction.
$f(W)$ Some function of body weight. Specifically used in

	Chapter 5 for the energy invested in each offspring.
F	Fecundity.
FMR	Field metabolic rate.
HRS	Home range size.
J	Rate of diffusion.
k	Constant.
k_1	Constant. See Equation (2.1).
k_2	Constant. See Equation (2.1).
k_3	Constant. See Equation (7.8).
k_4	Constant. See Equations (7.6) and (7.10).
k_5	Constant. See Equation (7.11).
k_6	Constant. See Equation (7.11).
k_7	Constant. See Equation (7.24).
k_8	Constant. See Equation (7.24).
K	Carrying capacity. Also used, in Chapter 6, to represent gross caloric content of food ingested, in Chapter 7 to represent k_3/k_1 and, in Chapter 9, to represent the permeability constant.
l	Length.
L	Some metric such as wing length or thorax length in *Drosophila melanogaster*.
LPI	Lifetime parental investment.
m	Constant. Also used, in Chapter 8, to represent male reproductive success.
M	Basal metabolism.
M_R	Moose's metabolism for maintenance, growth and reproduction.
n	Rank of male. Also used, in Chapter 8, to represent the position of a species on a resource axis.
n_1	Number of individuals. See Table 7.1.
n_2	Number of individuals with reproductive success not equal to 0. See Table 7.1.
n_3	Number of lumped classes. See Table 7.1.
ns	Not significant on a two-tailed test at the 5% level.
N	Number of points used to calculate a regression equation. Also used, in Chapter 3, as litter size, and, in Chapter 5, as the number of individuals.

NE Net energy intake by a moose.

p Two-tailed probability level. Also used, in Chapter 7, to represent the proportion of offspring sired by a dominant male and, in Chapter 9, to represent partial pressure.

p_e Partial pressure of oxygen in the environment.

p_o Partial pressure of oxygen at the centre of an animal.

P Productivity or production.

PI Parental investment.

PMR Mean metabolic rate when playing.

Q_L Heat loss.

r Pearson product-moment correlation coefficient. Also used, in Chapter 5, to represent the natural rate of increase and, in Chapter 9, to represent the radius of a spherical organism.

r_m Reproductive rate.

r_s Spearman rank correlation coefficient.

R Individual metabolic requirements. Also used, in Chapter 6, to represent daily rumen processing capacity.

REP Proportion of the energy budget used by an adult for reproduction in the absence of play.

RMR Resting metabolic rate.

s Increment in fitness to an adult 'mutant' playless individual. Also used, in Chapter 7, to represent speed and, in Chapter 9, to represent the distance over which oxygen is having to diffuse.

S Average number of individuals of a species that live in an area the size of the average home range of the species. Also used, in Chapter 6, to represent energetic cost of moving between food plants to crop a unit weight of food.

SE Standard error.

t Time. Specifically used in Chapter 8 to represent the proportion of time a male spends assimilating energy.

t_p Percentage of total time spent playing.

T Gestation time (for mammals) or incubation time (for birds).

T_a Age at maturity.

T_A Ambient temperature.

T_B Animal's temperature.

T_o Age at end of parental investment.

θ Exponent relating offspring weight at the end of parental investment interspecifically to W.

$\dot{V}_{O_2 max}$ Maximal rate of oxygen consumption.

W Body weight.

$\tilde{W}$ Body weight, when measured in different units to W.

W_f Adult female body weight.

W_{fl} That adult female body weight at which lifetime parental investment is maximized.

W_m Adult male body weight.

W_o Weight at end of parental investment. Also used, in Chapter 8, as the body weight of the lightest of a series of species arranged along a resource axis.

W_w Offspring weight at weaning.

x W^{a-b} in Chapter 5. Also used, in Chapter 5, for the reduction in k_1 for a hypothetical playless individual (See Equation (5.23)), in Chapter 8 for the ratio of the body weights of a pair of species arranged along a resource axis and, in Chapter 9, to represent the distance from the centre of a spherical organism.

X Size-related measure. See Equation (1.1).

y Dependent variable. Also used, in Chapter 5, to represent cx.

Y Size-related measure. See Equation (1.1).

References

Alberch, P., Gould, S.J., Oster, G.F. & Wake, D.B. (1979). Size and shape in ontogeny and phylogeny. *Paleobiology* **5,** 296–317.

Alexander, R.D., Hoogland, J.L., Howard, R.D., Noonan, K.M. & Sherman, P.W. (1979). Sexual dimorphism and breeding systems in pinnipeds, ungulates, primates and humans. In: *Evolutionary Biology and Human Social Behavior; An Anthropological Perspective*, Chagnon N.A. & Irons, W. (Eds.), Duxbury Press, North Scituate, Massachusetts, pp. 402–35.

Alexander, R.McN. (1971). *Size and Shape*, Edward Arnold, London.

Alexander, R.McN. (1977). Mechanics and scaling of terrestrial locomotion. In: *Scale Effects in Animal Locomotion*, Pedley, T.J. (Ed.), Academic Press, London, pp. 93–110.

Alexander, R.McN. (1979). *The Invertebrates*, Cambridge University Press, Cambridge.

Alexander, R.McN. (1982). *Optima for Animals*, Edward Arnold, London.

Alexander, R.McN., Jayes, A.S., Maloiy, G.M.O. & Wathuta, E.M. (1979). Allometry of the limb bones of mammals from shrews (*Sorex*) to elephants (*Loxodonta*). *J. Zool., Lond.* **189,** 305–14.

Altmann, J. (1980). *Baboon Mothers and Infants*, Harvard University Press, Cambridge, Massachusetts.

Altmann, S.A. & Wagner, S.S. (1978). A general model of optimal diet. In: *Recent Advances in Primatology: Volume 1 – Behaviour*, Chivers, D.J. & Herbert, J. (Eds.), Academic Press, London, pp. 407–14.

Anderson, S.S. & Fedak, M.A. (1985). Grey seal males: energetic and behavioural links between size and sexual success. *Anim. Behav.* **33,** 829–38.

Anderson, S.S., Burton, R.W. & Summers, C.F. (1975). Behaviour of grey seal (*Halichoerus grypus*) during a breeding season at North Rona. *J. Zool., Lond.* **177,** 179–95.

Andersson, E. (1969). Life cycle and growth of *Asellus aquaticus* (L.) with special reference to the effects of temperature. *Rep. Inst. Freshw. Res. Drottningholm* **49,** 5–26.

Apple, M.S. & Korostyshevskiy, M.A. (1980). Why many biological parameters are connected by power dependence. *J. theor. Biol.* **85,** 569–73.

Arak, A. (1983). Male–male competition and mate choice in anuran amphibians. In: *Mate Choice*, Bateson, P. (Ed.), Cambridge University Press, Cambridge, pp. 181–210.

Arman, P. & Hopcraft, D. (1975). Nutritional studies on East African herbivores. 1. Digestibilities of dry matter, crude fibre and crude protein in antelope, cattle and sheep. *Br. J. Nutr.* **33,** 255–64.

Atchley, W.R. (1978). Ratios, regression intercepts, and the scaling of data. *Syst. Zool* **27,** 78–83.

Baldock, B.M., Baker, J.M. & Sleigh, M.A. (1980). Laboratory growth rates of six species of freshwater Gymmamoebia. *Oecologia* (*Berl.*) **47,** 156–9.

Baldwin, N.S. (1956). Food consumption and growth of brook trout at different temperatures. *Trans. Am. Fish Soc.* **86,** 323–8.

Banks, C.J. (1964). Aphid nutrition and reproduction. *Rep. Rothamst. exp. Sta. for 1964*, pp. 299–309.

Banks, M.J. & Thompson, D.J. (1985). Lifetime mating success in the damselfly *Coenagrion puella. Anim. Behav.* **33,** 1175–83.

Banks, M.J. & Thompson, D.J. (1987). Lifetime reproductive success of the damselfly *Coenagrion puella. J. Anim. Ecol.* **56,** 815–32.

Banse, K. & Mosher, S. (1980). Adult body mass and annual production/biomass relationships of field populations. *Ecol. Monogr.* **50,** 355–79.

Barnes, R.D. (1974). *Invertebrate Zoology*, W.B. Saunders, Philadelphia.

Barrington, E.J.W. (1967). *Invertebrate Structure and Function*, Nelson, London.

Bartels, H. (1982). Metabolic rate of mammals equals the 0.75 power of their body weight. *Expl. Biol. Med.* **7,** 1–11.

Bartholomew, G.A. (1970). A model for the evolution of pinniped polygyny. *Evolution* **24,** 546–59.

Bartholomew, G.A. & Tucker, V. (1964). Size, body temperature, thermal conductance, oxygen consumption, and heart rate in Australian lizards. *Physiol. Zool.* **37,** 341–54.

Baur, M.E. & Friedl, R.R. (1980). Application of size-metabolism allometry to therapsids and dinosaurs. In: *A Cold Look at the Warm-Blooded Dinosaurs*, Thomas, R.D.K. & Olson, E.C. (Eds.), *AAAS Selected Symposium* **28,** 253–86.

Belovsky, G.E. (1978). Diet optimization in a generalist herbivore: the moose. *Theor. Pop. Biol.* **14,** 105–34.

Bennett, A.F. & Nagy, K.A. (1977). Energy expenditure in free-ranging lizards. *Ecology* **58,** 697–700.

Bennett, P.M. & Harvey, P.H. (1987). Active and resting metabolism in birds: allometry, phylogeny and ecology. *J. Zool., Lond.* **213,** 327–63.

Berry, J.F. & Shine, R. (1980). Sexual size dimorphism and sexual selection in turtles (Order Testudines). *Oecologia* (*Berl.*) **44,** 185–91.

Blair-West, J.R., Coghlan, J.P., Denton, D.A., Nelson, J.F., Orchard, E., Scoggins, B.A., Wright, R.D., Myers, K. & Junqueira, C.L. (1968). Physiological, morphological and behavioural adaptation to a sodium deficient environment by wild native Australian and introduced species of animals. *Nature* **217,** 922–8.

Blaxter, K.L. (1971). The comparative biology of lactation. In: *Lactation*, Falconer, I.R. (Ed.), Butterworths, London, pp. 51–69.

Blaxter, K.L., Kay, R.N.B., Sharman, G.A.M., Cunningham, J.M.M. & Hamilton, W.J. (1974). *Farming the Red Deer*, Her Majesty's Stationery Office, Edinburgh.

Blem, C.R. (1975). Energetics of nestling house sparrows *Passer domesticus*. *Comp. Biochem. Physiol.* **52A,** 305–12.

Bligh, J., Cloudsley-Thompson, J.L. & MacDonald, A.G. (1976). *Environmental Physiology of Animals*, Blackwell Scientific Publications, Oxford.

Blueweiss, L., Fox, H., Kudzma, V., Nakashima, D., Peters, R. & Sams, S. (1978). Relationships between body size and some life history parameters. *Oecologia* (*Berl.*) **37,** 257–72.

Blum, J.J. (1977). On the geometry of four-dimensions and the relationship between metabolism and body mass. *J. theor. Biol.* **64,** 599–601.

Boddington, M.J. (1978). An absolute metabolic scope for activity. *J. theor. Biol.* **75,** 443–9.

Boness, D.J. & James, H. (1979). Reproductive behaviour of the grey seal (*Halichoerus grypus*) on Sable Island, Nova Scotia. *J. Zool., Lond.* **188,** 477–500.

Bonner, J.T. (1965). *Size and Cycle: An Essay on the Structure of Biology*. Princeton University Press, Princeton, New Jersey.

Borgia, G. (1979). Sexual selection and the evolution of mating systems. In: *Sexual Selection and Reproductive Competition in Insects*, Blum, M. & Blum, A. (Eds.), Academic Press, New York, pp. 19–80.

Borgia, G. (1981). Mate selection in the fly *Scatophaga stercoraria*: female choice in a male-controlled system. *Anim. Behav.* **29,** 71–80.

Bradbury, J.W. (1977). Social organization and communication. In: *Biology of Bats: Volume III*, Wimsatt, W.A. (Ed.), Academic Press, New York, pp. 1–72.

Brett, J.R. (1979). Environmental factors and growth. In: *Fish Physiology, vol. VIII, Bioenergetics and Growth*, Hoar, W.S., Randall, D.J. & Brett, J.R. (Eds.), Academic Press, New York, pp. 599–675.

Brett, J.R. & Glass, N.R. (1973). Metabolic rates and critical swimming speeds of sockeye salmon (*Oncorhynchus nerka*) in relation to size and temperature. *J. Fish. Res. Bd. Can.* **30,** 379–87.

Brett, J.R. & Groves, T.D.D. (1979). Physiological energetics. In: *Fish Physiology, vol. VIII, Bioenergetics and Growth*, Hoar, W.S., Randall, D.J. & Brett, J.R. (Eds.), Academic Press, New York, pp. 279–352.

Brett, J.R. & Shelbourn, J.E. (1975). Growth rate of young sockeye salmon, *Oncorhynchus nerka*, in relation to fish size and ration level. *J. Fish. Res. Bd. Can.* **32,** 2103–10.

Brodie, P.F. (1975). Cetacean energetics, an overview of intraspecific size variation. *Ecology* **56,** 152–61.

Brody, S. (1945). *Bioenergetics and Growth: With Special Reference to the Efficiency Complex in Domestic Animals*, Reinhold Publishing Company, New York.

Brown, J.H. (1975). Geographical ecology of desert rodents. In: *Ecology and Evolution of Communities*, Cody, M.L. & Diamond, J.M. (Eds.), Belknap Press of Harvard University Press, Cambridge, Massachusetts, pp. 315–41.

Brown, J.H., Calder, W.A. III & Kodric-Brown, A. (1978). Correlates and consequences of body size in nectar-feeding birds. *Amer. Zool.* **18,** 687–700.

Brown, L. (1981). Patterns of female choice in mottled sculpins (Cottidae, Teleostei). *Anim. Behav.* **29,** 375–82.

Brown, M.E. (1946). The growth of brown trout (*Salmo trutta* Linn.). II. The growth of two-year-old trout at a constant temperature of 11.5 °C. *J. Exp. Biol.* **22,** 118–29.

Bryant, D.M. & Gardiner, A. (1979). Energetics of growth in house martins (*Delichon urbica*). *J. Zool., Lond.* **189,** 275–304.

Bryden, M.M. (1969). Growth of the southern elephant seal, *Mirounga leonina* (Linn.). *Growth* **33,** 531–6.

Bucher, T.L., Ryan, M.J. & Bartholomew, G.A. (1982). Oxygen consumption during resting, calling, and nest building in the frog *Physalaemus pustulosus*. *Physiol. Zool.* **55,** 10–22.

Cain, B.W. (1976). Energetics of growth for black-bellied tree ducks. *Condor* **78,** 124–8.

Calder, W.A. III (1974). Consequences of body size for avian energetics. In: *Avian Energetics*, Paynter, R.A. Jr. (Ed.), Publications of the Nuttall Ornithological Club, no. 15, Cambridge, Massachusetts, pp. 86–144.

Calder, W.A. III (1984). *Size, Function and Life History*. Harvard University Press, Cambridge, Massachusetts.

Calow, P. (1977). Conversion efficiencies in heterotrophic organisms. *Biol. Rev.* **52,** 385–409.

Calow, P. (1979). The cost of reproduction – a physiological approach. *Biol. Rev.* **54,** 23–40.

Cammen, L.M. (1980). Ingestion rate: an empirical model for aquatic deposit feeders and detritivores. *Oecologia* (*Berl.*) **44,** 303–10.

Carey, F.G. (1982). Warm fish. In: *A Companion to Animal Physiology*, Taylor, C.R., Johansen, K. & Bolis, L. (Eds.), Cambridge University Press, Cambridge, pp. 216–33.

Carne, P.B. (1966). Growth and food consumption during the larval stages of *Paropsis atomaria* (Coleoptera: Chrysomelidae). *Ent. exp. & appl.* **9,** 105–12.

Case, T.J. (1978a). On the evolution and adaptive significance of postnatal growth rates in the terrestrial vertebrates. *Q. Rev. Biol.* **53,** 243–82.

Case, T.J. (1978b). Endothermy and parental care in the terrestrial vertebrates. *Am. Nat.* **112,** 861–74.

Casey, T.M. (1981). A comparison of mechanical and energetic flight cost for hovering sphinx moths. *J. exp. Biol.* **91,** 117–29.

Chaplin, S.B. & Chaplin, S.J. (1981). Comparative growth energetics of a migratory and nonmigratory insect: the milkweed bugs. *J. Anim. Ecol.* **50,** 407–20.

Chesney, E.J. Jr & Estevez, J.I. (1976). Energetics of winter flounder (*Pseudopleuronectes americanus*) fed the polychaete, *Nereis virens*, under experimental conditions. *Trans. Am. Fish Soc.* **105,** 592–5.

Clark, A.B. (1978). Sex ratio and local resource competition in a prosimian primate. *Science* **201,** 163–5.

Clements, E.T. & Stevens, C.E. (1980). A comparison of gastrointestinal transit time in ten species of mammal. *J. agric. Sci., Camb.* **94,** 735–7.

Clutton-Brock, T.H. (1984). Reproductive effort and terminal investment in iteroparous animals. *Am. Nat.* **123,** 212–29.

Clutton-Brock, T.H. (Ed.) (1988). *Reproductive Success: Studies of Individual Variation in Contrasting Breeding Systems*, Chicago University Press, Chicago.

Clutton-Brock, T.H. & Albon, S.D. (1982). Parental investment and the sex ratio of progeny in mammals. In: *Current Problems in Sociobiology*, King's College Sociobiology Group (Ed.), Cambridge University Press, Cambridge, pp. 223–47.

Clutton-Brock, T.H. & Harvey, P.H. (1976). Evolutionary rules and primate societies. In: *Growing Points in Ethology*, Bateson, P.P.G. & Hinde, R.A. (Eds.), Cambridge University Press, Cambridge, pp. 195–237.

Clutton-Brock, T.H. & Harvey, P.H. (1977). Primate ecology and social organization. *J. Zool., Lond.* **183,** 1–39.

Clutton-Brock, T.H. & Harvey, P.H. (1979). Comparison and adaptation. *Proc. R. Soc. Lond. B* **205,** 547–65.

Clutton-Brock, T.H. & Harvey, P.H. (1983). The functional significance of variation in body size among mammals. In: *Advances in the Study of Mammalian Behavior*, Eisenberg, J.F. & Kleiman, D.G. (Eds.), *Spec. Publ. Amer. Soc. Mamm.* **7,** 632–63.

Clutton-Brock, T.H., Albon, S.D. & Harvey, P.H. (1980). Antlers, body size and breeding group size in the Cervidae. *Nature* **285,** 565–7.

Clutton-Brock, T.H., Harvey, P.H. & Rudder, B. (1977). Sexual dimorphism, socionomic sex ratio and body weight in primates. *Nature* **269,** 797–800.

Clutton-Brock, T.H., Albon, S.D., Gibson, R.M. & Guinness, F.E. (1979). The logical stage: adaptive aspects of fighting in red deer (*Cervus elaphus* L.). *Anim. Behav.* **27,** 211–25.

Convey, P. (1987). *Influences on Mating Behaviour and Reproductive Success in the Odonata*, Ph.D. thesis, University of Cambridge.

Convey, P. (In prep.). Influences on the choice between territorial and satellite behaviour in male *Libellula quadrimaculata* Linn. (Odonata: Libellulidae).

Corbet, G.B. & Southern, H.N. (Eds.) (1977). *The Handbook of British Mammals*, 2nd edn, Blackwell Scientific Publications, Oxford.

Cornet, A. & Jouventin, P. (1980). Le phoque de Weddell (*Leptonychotes weddelli* L.) à Pointe Géologie et sa plasticité sociale. *Mammalia* **44,** 497–521.

Crompton, A.W., Taylor, C.R. & Jagger, J.A. (1978). Evolution of homeothermy in mammals. *Nature* **272,** 333–6.

Crook, J.H. (1962). The adaptive significance of pair formation types in weaver birds. *Symp. zool. Soc. Lond.* **8,** 57–70.

Croxall, J.P. (1982). Energy costs of incubation and moult in petrels and penguins. *J. Anim. Ecol.* **51,** 177–94.

Crump, M.L. (1974). Reproductive strategies in a tropical anuran community. *University of Kansas Museum of Natural History Miscellaneous Publication*

No. 61, 1–68.
Daborn, G.R. (1975). Life history and energy relations of the giant fairy shrimp *Branchinecta gigas* Lynch 1937 (Crustacea: Anostraca). *Ecology* **56,** 1025–39.
Damuth, J. (1981a). Population density and body size in mammals. *Nature* **290,** 699–700.
Damuth, J. (1981b). Home range, home range overlap, and species energy use among herbivorous mammals. *Biol. J. Linn. Soc.* **15,** 185–193.
Darwin, C. (1871). *The Descent of Man and Selection in Relation to Sex*, John Murray, London.
Davies, N.B. & Halliday, T.R. (1977). Optimal mate selection in the toad *Bufo bufo*. *Nature* **269,** 56–8.
Davies, N.B. & Halliday, T.R. (1978). Deep croaks and fighting assessment in toads *Bufo bufo*. *Nature* **274,** 683–5.
Davies, N.B. & Halliday, T.R. (1979). Comparative mate searching in male common toads, *Bufo bufo*. *Anim. Behav.* **27,** 1253–67.
Dawes, B. (1930a). Growth and maintenance in the plaice. Part I. *J. Marine Biol. Ass. N. S.* **17,** 103–74.
Dawes, B. (1930b). Growth and maintenance in the plaice. Part II. *J. Marine Biol. Ass. N. S.* **17,** 877–947.
Demment, M.W. (1982). The scaling of ruminoreticulum size with body weight in East African ungulates. *Afr. J. Ecol.* **20,** 43–7.
Demment, M.W. (1983). Feeding ecology and the evolution of body size of baboons. *Afr. J. Ecol.* **21,** 219–33.
Demment, M.V. & Van Soest, P.J. (1985). A nutritional explanation for body-size patterns of ruminant and nonruminant herbivores. *Am. Nat.* **125,** 641–72.
Dewar, A.M. (1977). Assessment of methods for testing varietal resistance to aphids in cereals. *Ann. appl. Biol.* **87,** 183–90.
Diamond, J.M. (1975). Assembly of species communities. In: *Ecology and Evolution of Communities*, Cody, M.L. & Diamond, J.M. (Eds.), Belknap Press of Harvard University Press, Cambridge, Massachusetts, pp. 342–444.
Dittus, W.P.J. (1979). The evolution of behaviors regulating density and age-specific sex ratios in a primate population. *Behaviour* **69,** 265–302.
Dixon, A.F.G. (1970). Quality and availability of food for a sycamore aphid population. In: *Animal Populations in Relation to their Food Resources*, Watson, A. (Ed.), Blackwell Scientific Publications, Oxford, pp. 271–86.
Dixon, A.F.G. (1971). The role of intra-specific mechanisms and predation in regulating the numbers of the lime aphid, *Eucallipterus tiliae* L.. *Oecologia* (*Berl.*) **8,** 179–93.
Dixon, A.F.G. (1976). Reproductive strategies of the alate morphs of the bird cherry-oat aphid *Rhopalosiphum padi* L.. *J. Anim. Ecol.* **45,** 817–30.
Dixon, A.F.G. & Dharma, T.R. (1980a). Number of ovarioles and fecundity in the black bean aphid, *Aphis fabae*. *Ent. exp. & appl.* **28,** 1–14.
Dixon, A.F.G. & Dharma, T.R. (1980b). 'Spreading of the risk' in developmental mortality: size, fecundity and reproductive rate in the black bean aphid. *Ent. exp. & appl.* **28,** 301–12.

Dixon, A.F.G. & Wratten, S.D. (1971). Laboratory studies on aggregation, size and fecundity in the black bean aphid, *Aphis fabae* Scop. *Bull. ent. Res.* **61,** 97–111.

Dixson, A.F. (1987). Observations on the evolution of the genitalia and copulatory behaviour in male primates. *J. Zool., Lond.* **213,** 423–43.

Donhoffer, S. (1986). Body size and metabolic rate: exponent and coefficient of the allometric equation: the role of units. *J. theor. Biol.* **119,** 125–37.

Downhower, J.F. (1976). Darwin's finches and the evolution of sexual dimorphism in body size. *Nature* **263,** 558–63.

Drożdż, A., Górecki, A., Grodziński, W. & Pelikán, J. (1971). Bioenergetics of water voles (*Arvicola terrestris* L.) from southern Moravia. *Ann. Zool. Fennici* **8**, 97–103.

Dunger, W. (1958). Über die Zersetzung der Laubstreu durch die Boden-Makrofauna in Auenwald. *Zoologische Jahrbücher Abteilung für systematik ökologie und geographie der tiere* **86,** 139–80.

Dunham, A.E., Tinkle, D.W. & Gibbons, J.W. (1978). Body size in lizards: a cautionary tale. *Ecology* **59,** 1230–8.

Earhart, C.M. & Johnson, N.K. (1970). Size dimorphism and food habits of North American owls. *Condor* **72,** 251–64.

Economos, A.C. (1979a). On structural theories of basal metabolic rate. *J. theor. Biol.* **80,** 445–50.

Economos, A.C. (1979b). Gravity, metabolic rate and body size of mammals. *The Physiologist* **22,** S-71–S-72.

Economos, A.C. (1981). The largest land mammal. *J. theor. Biol.* **89,** 211–5.

Egorov, O.V. (1964). *Wild ungulates of Yakutia, Izdatel'stvo 'Nauko'* [translated from the Russian by the Israel Program for Scientific Translations, Jerusalem, 1967].

Eisenberg, J.F. & Wilson, D.E. (1978). Relative brain size and feeding strategies in the Chiroptera. *Evolution* **32,** 740–51.

Elliott, G.F. (1948). The evolutionary significance of brachial development in terebratelloid brachiopods. *Ann. Mag. nat. Hist.* **I** (12th ser.), 297–317.

Elliot, J.M. (1979). Energetics of freshwater teleosts. *Symp. zool. Soc. Lond.* **44,** 29–61.

Ellis, R.J. (1971). Notes on the biology of the isopod *Asellus tomalensis* Harford in an intermittent pond. *Trans. Amer. Micros. Soc.* **90,** 51–61.

Eltringham, S.K. (1979). *The Ecology and Conservation of Large African Mammals*, MacMillan, London.

Epp, R.W. & Lewis, W.M. Jr. (1980). The nature and ecological significance of metabolic changes during the life history of copepods. *Ecology* **61,** 259–64.

Erlinge, S. (1979). Adaptive significance of sexual dimorphism in weasels. *Oikos* **33,** 233–45.

Evans, D.E. (1962). The food requirements of *Phonoctonus nigrofasciatus* Ståhl. (Hemiptera, Reduviidae). *Ent. exp. & appl.* **5,** 33–9.

Fagen, R.M. (1976). Exercise, play, and physical training in animals. In: *Perspectives in Ethology, Vol. 2*, Bateson, P.P.G. & Klopfer, P.H. (Eds.), Plenum Press, New York, pp. 189–219.

Fagen, R.M. (1977). Selection for optimal age-dependent schedules of play

behavior. *Am. Nat.* **111,** 395–414.
Fagen, R. (1981). *Animal Play Behavior.* Oxford University Press, New York.
Farlow, J.O. (1976). A consideration of the trophic dynamics of a Late Cretaceous large-dinosaur community (Oldman Formation). *Ecology* **57,** 841–57.
Fedak, M.A. & Seeherman, H.J. (1979). Reappraisal of energetics of locomotion shows identical cost in bipeds and quadrupeds including ostrich and horse. *Nature* **282,** 713–16.
Feldman, H.A. & McMahon, T.A. (1983). The $\frac{3}{4}$ mass exponent for energy metabolism is not a statistical artifact. *Respir. Physiol.* **52,** 149–63.
Fenchel, T. (1974). Intrinsic rate of natural increase: the relationship with body size. *Oecologia (Berl.)* **14,** 317–26.
Ferns, P.N. (1979). Food consumption and energy expenditure of the field vole in the laboratory and in a small outdoor enclosure. *Acta theriol.* **24,** 47–59.
Fewkes, D.W. (1960). The food requirements by weight of some British Nabidae (Heteroptera). *Ent. exp. & appl.* **3,** 231–7.
Fincke, O.M. (1982). Lifetime mating success in a natural population of the damselfly, *Enallagma hageni* (Walsh) (Odonata: Coenagrionidae). *Behav. Ecol. Sociobiol.* **10,** 293–302.
Finlay, B.J. (1977). The dependence of reproductive rate on cell size and temperature in freshwater ciliated Protozoa. *Oecologia (Berl.)* **30,** 75–81.
Fischer, Z. (1972). The energy budget of *Lestes sponsa* (Hans.) during its larval development. *Pol. Arch. Hydrobiol.* **19,** 215–22.
Fleming, T.H. (1975). The role of small mammals in tropical ecosystems. In: *Small Mammals: Their Productivity and Population Dynamics*, Golley, F.B., Petrusewicz, K. & Ryszkowski, L. (Eds.), Cambridge University Press, Cambridge, pp. 269–98.
Forbes, J.M. (1982). Prediction of the voluntary intake of complete feeds by growing cattle. *British Society of Animal Production Winter Meeting*, Paper No. 38.
Frazer, J.F.D. & Huggett, A.St.G. (1973). Specific foetal growth rates of cetaceans. *J. Zool., Lond.* **169,** 111–26.
Frazer, J.F.D. & Huggett, A.St.G. (1974). Species variations in the foetal growth rates of eutherian mammals. *J. Zool., Lond.* **174,** 481–509.
French, N.R., Grant, W.R., Grodziński, W. & Swift, D.M. (1976). Small mammal energetics in grassland ecosystems. *Ecol. Monogr.* **46,** 201–20.
Fry, F.E.J. (1971). The effect of environmental factors on the physiology of fish. In: *Fish Physiology, volume VI, Environmental Relations and Behaviour*, Hoar, W.S. & Randall, D.J. (Eds.), Academic Press, New York, pp. 1–98.
Gatz, A.J. (1981). Non-random mating by size in American toads, *Bufo americanus. Anim. Behav.* **27,** 1253–67.
Gause, G.F. (1934). *The Struggle for Existence*, Hafner, New York.
Gehr, P., Mwangi, D.K., Ammann, A., Maloiy, G.M.O., Taylor, C.R. & Weibel, E.R. (1981). Design of the mammalian respiratory system. V. Scaling morphometric pulmonary diffusing capacity to body mass: wild and domestic mammals. *Respir. Physiol.* **44,** 61–86.
Geist, V. (1971). *Mountain Sheep: A Study in Behavior and Evolution*, University of Chicago Press, Chicago.

Geist, V. (1987). Bergmann's rule is invalid. *Can. J. Zool.* **65,** 1035–8.

Gentry, R.L. (1973). Thermoregulatory behaviour of eared seals. *Behaviour* **46,** 73–93.

Gentry, R.L. (1975). Comparative social behaviour of eared seals. *Rapp. P.-v. Réun. Cons. Perm. int. Explor. Mer.* **169,** 188–94.

Gerald, V.M. (1976). The effect of size on the consumption absorption and conversion of food in *Ophiocephalus punctatus* Bloch. *Hydrobiologia* **49,** 77–85.

Gere, G. (1956). The examination of the feeding biology and the humificative function of Diploda and Isopoda. *Acta Biol. Hung.* **6,** 257–71.

Gerking, S.D. (1952). The protein metabolism of sunfishes of different ages. *Physiol. Zool.* **25,** 358–72.

Gessaman, J.A. (1973). Methods of estimating the energy cost of free existence. In: *Ecological Energetics of Homeotherms: A View Compatible with Ecological Modeling*, Gessaman, J.A. (Ed.), Utah State University Press, Logan, Utah, pp. 3–31.

Ghiselin, M.T. (1974). *The Economy of Nature and the Evolution of Sex*, University of California Press, Berkeley.

Gibb, J. (1954). Feeding ecology of tits, with notes on treecreeper and goldcrest. *Ibis* **96,** 513–43.

Gibson, R.M. (1978). *Behavioural Factors Affecting Male Reproductive Success in Red Deer Stags*, Doctor of Philosophy thesis, University of Sussex.

Gibson, R.M. & Guinness, F.E. (1980). Differential reproduction among red deer (*Cervus elaphus*) stags on Rhum. *J. Anim. Ecol.* **49,** 199–208.

Gilbert, F.S. (1982). *Morphology and the Foraging Ecology of Hoverflies (Diptera: Syrphidae)*, Doctor of Philosophy thesis, University of Cambridge.

Glass, N.R. (1969). Discussion of calculation of power function with special reference to respiratory metabolism in fish. *J. Fish. Res. Bd. Can.* **26,** 2643–50.

Godfrey, N.W. (1961a). The functional development of the calf I. Growth of the stomach of the calf. *J. Agric. Sci.* **57,** 173–5.

Godfrey, N.W. (1961b). The functional development of the calf II. Development of rumen function in the calf. *J. Agric. Sci.* **57,** 177–83.

Golley, F.B. (1961). Energy values of ecological materials. *Ecology* **42,** 581–4.

Górecki, A. (1971). Metabolism and energy budget in the harvest mouse. *Acta theriol.* **16,** 213–20.

Górecki, A. (1977). Energy flow through the common hamster population. *Acta theriol.* **22,** 25–66.

Gould, S.J. (1966). Allometry and size in ontogeny and phylogeny. *Biol. Rev.* **41,** 587–640.

Gould, S.J. (1975a). Allometry in primates, with emphasis on scaling and the evolution of the brain. *Contrib. Primat.* **5,** 244–92.

Gould, S.J. (1975b). On the scaling of tooth size in mammals. *Amer. Zool.* **15,** 351–62.

Gould, S.J. (1978). Morton's ranking of races by cranial capacity. *Science* **178,** 503–9.

Gould, S.J. & Lewontin, R.C. (1979). The spandrels of San Marco and the

Panglossian paradigm: a critique of the adapationist programme. *Proc. R. Soc. Lond. B* **205,** 581–98.

Grant, S.A. & Campbell, D.R. (1978). Seasonal variation in in vitro digestibility and structural carbohydrate content of some commonly grazed plants of blanket bog. *J. Brit. Grassland Soc.* **31,** 167–73.

Green, N.P.O., Stout, G.W. & Taylor, D.J. (1984). *Biological Science 1: Organisms, Energy and Environment*, Cambridge University Press, Cambridge.

Greenspan, B.V. (1980). Male size and reproductive success in the communal courtship system of the fiddler crab *Uca rapax*. *Anim. Behav.* **28,** 387–92.

Greenstone, M.H. (1979). Spider feeding behaviour optimises dietary essential amino acid composition. *Nature* **282,** 501–3.

Grodziński, W. (1967). Daily metabolism rate and body size of common voles *Microtus arvalis* Pall. *Small Mammal Newsletter* **1(3),** 5–6.

Grodziński, W. (1971). Energy flow through populations of small mammals in the Alaskan Taiga forest. *Acta theriol.* **16,** 231–75.

Grodziński, W. & Górecki, A. (1967). Daily energy budgets of small rodents. In: *Secondary Productivity of Terrestrial Ecosystems* (*Principles and Methods*) *vol. 1*, Petrusewicz, K. (Ed.), Pánstwowe wydawnictwo naukowe, Warszawa, pp. 295–314.

Günther, B. (1975). Dimensional analysis and theory of biological similarity. *Physiol. Rev.* **55,** 659–99.

Günther, B. & Léon de la Barra, B. (1966). Physiometry of the mammalian circulatory system. *Acta Physiol. Latinoam* **16,** 32–42.

Günther, B. & Morgado, E. (1982). Theory of biological similarity revisited. *J. theor. Biol.* **96,** 543–59.

Hainsworth, F.R. (1973). On the tongue of a hummingbird: its role in the rate and energetics of feeding. *Comp. Biochem. Physiol.* **46A,** 65–78.

Hainsworth, F.R. & Wolf, L.L. (1972a). Power for hovering flight in relation to body size in hummingbirds. *Am. Nat.* **106,** 589–96.

Hainsworth, F.R. & Wolf, L.L. (1972b). Crop volume, nectar concentration and hummingbird energetics. *Comp. Biochem. Physiol.* **42A,** 859–66.

Hamilton, W.D. (1967). Extraordinary sex ratios. *Science* **156,** 477–88.

Hansson, L. & Grodziński, W. (1970). Bioenergetic parameters of the field vole *Microtus agrestis* L. *Oikos* **21,** 76–82.

Hanwell, A. & Peaker, M. (1977). Physiological effects of lactation on the mother. *Symp. zool. Soc. Lond.* **41,** 297–312.

Harcourt, A.H., Harvey, P.H., Larson, S.G. & Short, R.V. (1981). Testis weight, body weight and breeding system in primates. *Nature* **293,** 55–7.

Harestad, A.S. & Bunnell, F.L. (1979). Home range and body weight – a reevaluation. *Ecology* **60,** 389–402.

Harper, J.L. (1977). *Population Biology of Plants*, Academic Press, London.

Harvey, P.H. & Bennett, T.M. (1985). Sexual dimorphism and reproductive strategies. In: *Human Sexual Dimorphism*, Ghesquiere, J., Martin, R.D. & Newcombe, F. (Eds.), Taylor & Francis, London, pp. 43–59.

Harvey, P.H. & Mace, G.M. (1982). Comparisons between taxa and adaptive trends: problems of methodology. In: *Current Problems in Sociobiology*, King's College Sociobiology Group, Cambridge (Ed.), Cambridge University Press, Cambridge, pp. 343–61.

Harvey, P.H., Clutton-Brock, T.H. & Mace, G.M. (1980). Brain size and ecology in small mammals and primates. *Proc. Nat. Acad. Sci. USA* **77,** 4387–9.

Harvey P.H., Kavanagh, M. & Clutton-Brock, T.H. (1978a). Canine tooth size in female primates. *Nature* **276,** 817–8.

Harvey, P.H., Kavanagh, M. & Clutton-Brock, T.H. (1978b). Sexual dimorphism in primate teeth. *J. Zool., Lond.* **186,** 475–85.

Hatanaka, M., Kosaka, M. & Satô, Y. (1956a). Growth and food consumption in plaice. Part I. *Limanda yokohamae* (Günther). *Tohoku J. Agr. Res.* **7,** 151–62.

Hatanaka, M., Kosaka, M. & Satô, Y. (1956b). Growth and food consumption in plaice. Part II. *Kareius bicoloratus* (Basilewsky). *Tohoku J. Agr. Res.* **7**.

Havlicek, L.L. & Peterson, N.L. (1976). Robustness of the Pearson correlation against violations of the assumptions. *Perceptual and Motor Skills* **43,** 1319–34.

Heck, K.L. Jr. (1976). Some critical considerations of the theory of species packing. *Evol. Theory* **1,** 247–58.

Heinroth, O. (1922). Die Beziehungen zwischen Vogelgewicht, Eigewicht, Gelegegewicht und Brutdauer. *J. Ornithol.* **70,** 172–285.

Hemmingsen, A.M. (1960). Energy metabolism as related to body size and respiratory surfaces, and its evolution. *Rep. Steno Mem. Hosp. Nord. Insul.* **9(2),** 7–110.

Herreid, C.F. II & Kessel, B. (1967). Thermal conductance in birds and mammals. *Comp. Biochem. Physiol.* **21,** 405–14.

Heusner, A.A. (1982). Energy metabolism and body size. I. Is the 0.75 mass exponent of Kleiber's equation a statistical artifact? *Respir. Physiol.* **48,** 1–12.

Hill, A.V. (1950). The dimensions of animals and their muscular dynamics. *Sci. Progr., London* **38,** 209–30.

Hill, G. & Holman, J. (1986). *Science 1*, Nelson, Walton-on-Thames, Surrey.

Hinde, R.A. (1952). *The behaviour of the great tit* (Parus major) *and some other related species.* Behaviour supplement II, E.J. Brill, Leiden.

Hiratsuka, E. (1920). Researches on the nutrition of the silk worm. *Bull. ser. Exp. Sta. Japan* **1,** 257–315.

Hirshfield, M.F. & Tinkle, D.W. (1975). Natural selection and the evolution of reproductive effort. *Proc. Nat. Acad. Sci. USA* **72,** 2227–31.

Hoppe, P.P. (1977). Rumen fermentation and body weight in African ruminants. *Proc. 13th Congr. Game Biol.*, Atlanta, pp. 141–50.

Howard, R.D. (1978a). The influence of male-defended oviposition sites on early embryo mortality in bullfrogs. *Ecology* **59,** 789–98.

Howard, R.D. (1978b). The evolution of mating strategies in bullfrogs, *Rana catesbeiana. Evolution* **32,** 850–71.

Howard, R.D. (1979). Estimating reproductive success in natural populations. *Am. Nat* **114,** 221–31.

Howard, R.D. (1980). Mating behaviour and mating success in woodfrogs, *Rana sylvatica*. *Anim. Behav.* **28,** 705–16.
Hozumi, K. (1985). Phase diagrammatic approach to the analysis of growth curve using the u–w diagram – basic aspects. *Bot. mag. Tokyo* **98,** 239–50.
Hubbell, S.P. (1980). Seed predation and the coexistence of tree species in tropical forests. *Oikos* **35,** 214–29.
Hubbs, C., Stevenson, M.M. & Peden, A.E. (1968). Fecundity and egg size in two central Texas darter populations. *Southwest. Nat.* **13,** 301–24.
Huggett, A.St G. & Widdas, W.F. (1951). The relationship between mammalian foetal weight and conception age. *J. Physiol.* **114,** 306–17.
Humphreys, W.F. (1979). Production and respiration in animal communities. *J. Anim. Ecol.* **48,** 427–53.
Hunt, R. (1978). *Plant Growth Analysis*, Edward Arnold, London.
Hunt, R. & Parsons, I.T. (1974). A computer program for deriving growth-functions in plant growth-analysis. *J. appl. Ecol.* **11,** 297–307.
Hurd, R.G. (1977). Vegetative plant growth analysis in controlled environments. *Ann. Bot.* **41,** 779–87.
Hutchinson, G.E. (1959). Homage to Santa Rosalia or Why are there so many kinds of animals? *Am. Nat.* **93,** 145–59.
Huxley, J.S. (1927). On the relation between egg-weight and body-weight in birds. *J. Linn. Soc., Zool.* **36,** 457–66.
Huxley, J.S. (1932). *Problems of Relative Growth*, Methuen, London.
Iberall, A.S. (1973). On growth, form, and function – a fantasia on the design of a mammal. *Journal of Dynamic Systems, Measurement and Control*, pp. 291–5.
Ikeda, T. (1977). Feeding rates of planktonic copepods from a tropical sea. *J. exp. mar. Biol. Ecol.* **29,** 263–77.
Jarman, P.J. (1974). The social organisation of antelope in relation to their ecology. *Behav.* **48,** 215–67.
Jerison, H.J. (1973). *Evolution of the Brain and Intelligence*, Academic Press, New York.
Jobling, M., Gwyther, D. & Grove, D.J. (1977). Some effects of temperature, meal size and body weight on gastric evacuation time in the dab *Limanda limanda* (L.). *J. Fish Biol.* **10,** 291–8.
Johnson, C.G. (1960). The relation of weight of food ingested to increase in bodyweight during growth in the bed-bug, *Cimex lectularius* L. (Hemiptera). *Ent. exp. & appl.* **3,** 238–40.
Jones, G. & Jones, M. (1984). *Biology: A Course to 16+*, Cambridge University Press, Cambridge.
Jordan, C.F. (1971). A world pattern in plant energetics. *Amer. Sci.* **59,** 425–33.
Jouventin, P. & Cornet, A. (1979). La vie sociale des phoques. *La Recherche* **10,** 1058–66.
Kaplan, R.H. & Salthe, S.N. (1979). The allometry of reproduction: an empirical view in salamanders. *Am. Nat.* **113,** 671–89.
Karasov, W.H. (1986). Energetics, physiology and vertebrate ecology. *TREE* **1,** 101–4.
Kaufman, G.W., Siniff, D.B. & Reichle, R. (1975). Colony behaviour of

Weddell seals, *Leptonychotes weddelli*, at Hutton Cliffs, Antarctica. *Rapp. P.-v. Reun. Cons. int. Explor. Mer* **169,** 228–46.

Kempton, R.A., Lowe, H.J.B. & Bintcliffe, E.J.B. (1980). The relationship between fecundity and adult weight in *Myzus persicae*. *J. Anim. Ecol.* **49,** 917–26.

Kenagy, G.J. & Trombulak, S.C. (1986). Size and function of mammalian testes in relation to body size. *J. Mamm.* **67,** 1–22.

Kendeigh, S.C. (1972). Energy control of size limits in birds. *Am. Nat.* **106,** 79–88.

Kendeigh, S.C., Dol'nik, V.R. & Gavrilov, V.M. (1977). Avian energetics. In: *Granivorous Birds in Ecosystems*, Pinowski, J. & Kendeigh, S.C. (Eds.), Cambridge University Press, Cambridge, pp. 127–204.

Kermack, K.A. & Haldane, J.B.S. (1950). Organic correlation and allometry. *Biometrika* **37,** 30–41.

Kidwell, J.F. & Chase, H.B. (1967). Fitting the allometric equation – a comparison of ten methods by computer simulation. *Growth* **31,** 165–79.

King, J.R. (1974). Seasonal allocation of time and energy resources in birds. In: *Avian energetics*, Paynter, R.A. Jr. (Ed.), Publications of the Nuttall Ornithological Club, Cambridge, Massachusetts, pp. 4–70.

Kinne, O. (1960). Growth, food intake, and food conversion in a euryplastic fish exposed to different temperatures and salinities. *Physiol. Zool.* **33,** 288–317.

Kleiber, M. (1947). Body size and metabolic rate. *Physiol. Rev.* **27,** 511–41.

Kleiber, M. (1961). *The Fire of Life*, John Wiley, London.

Kleiber, M. (1972). Body size, conductance for animal heat flow and Newton's law of cooling. *J. theor. Biol.* **37,** 139–50.

Kleiber, M. (1975). *The Fire of Life: An Introduction to Animal Energetics*, Robert E. Krieger, Huntingdon, New York.

Kleiman, D.G. (1977). Monogamy in mammals. *Q. Rev. Biol.* **52,** 39–69.

Klein, D.R. (1964). Range-related differences in growth of deer reflected in skeletal ratios. *J. Mamm.* **45,** 226–35.

Klekowski, R.Z., Schiemer, F. & Duncan, A. (1979). A bioenergetics study of a benthic nematode, *Plectus palustris* de Man 1880, throughout its life cycle. I. The respiratory metabolism at different densities of bacterial food. *Oecologia* (*Berl.*) **44,** 119–24.

Kohler, A.C. (1964). Variations in the growth of Atlantic cod (*Gadus morhua* L.). *J. Fish. Res. Bd. Can.* **21,** 57–100.

Koller, C.N. & Leonard, D.E. (1981). Comparison of energy budgets for spruce budworm *Choristoneura fumiferana* (Clemens) on balsam fir and white spruce. *Oecologia* (*Berl.*) **49,** 14–20.

Krebs, C.J. (1972). *Ecology: The Experimental Analysis of Distribution and Abundance*, Harper & Row, New York.

Krebs, J.R. & Davies, N.B. (1987). *An Introduction to Behavioural Ecology*, 2nd ed., Blackwell Scientific Publications, Oxford.

Krebs, J.R. & McCleery, R.H. (1984). Optimization in behavioural ecology. In: *Behavioural Ecology: An Evolutionary Approach*, 2nd ed., Krebs, J.R. & Davies, N.B. (Eds.), Blackwell Scientific Publications, Oxford, pp. 91–121.

Lack, D. (1968). *Ecological Adaptations for Breeding in Birds*, Methuen, London.

Laird, A.K. (1965). Dynamics of relative growth. *Growth* **29,** 249–63.

Lampert, W. (1977). Studies of the carbon balance of *Daphnia pulex* de Geer as related to environmental conditions. II. The dependence of carbon assimilation on animal size, food concentration and diet species. *Arch. Hydrobiol./Suppl.* **48,** 310–35.

Lasiewski, R.C., Weathers, W.W. & Bernstein, M.H. (1967). Physiological responses of the giant hummingbird, *Patagona gigas. Comp. Biochem. Physiol.* **23,** 797–813.

Lavigne, D.M. (1982). Similarity in energy budgets of animal populations. *J. Anim. Ecol.* **51,** 195–206.

Lawton, J.H. (1981). Moose, wolves, *Daphnia*, and *Hydra*: on the ecological efficiency of endotherms and ectotherms. *Am. Nat.* **117,** 782–3.

Laybourn-Parry, J. & Strachan, I.M. (1980). Respiratory metabolism of *Cyclops bicuspidatus* (*sensu stricta*) (Claus) (Copepoda: Cyclopoida) from Esthwaite Water, Cumbria. *Oecologia* (*Berl.*) **46,** 386–90.

La Boeuf, B.J. (1974). Male–male competition and reproductive success in elephant seals. *Am. Nat.* **14,** 163–76.

Ledger, H.P. (1968). Body composition as a basis for a comparative study of some East African mammals. *Symp. zool. Soc. Lond.* **21,** 289–310.

Leitch, I., Hytten, F.E. & Billewicz, W.Z. (1959). The maternal and neonatal weights of some Mammalia. *Proc. zool. Soc. Lond.* **133,** 11–28.

Leutenegger, W. (1973). Maternal-fetal weight relationships in primates. *Folia primat.* **20,** 280–93.

Leutenegger, W. (1976). Allometry of neonatal size in eutherian mammals. *Nature* **263,** 229–30.

Leutenegger, W. (1978). Scaling of sexual dimorphism in body size and breeding system in primates. *Nature* **272,** 610–1.

Leuthold, W. (1977). *African Ungulates: A Comparative Review of their Ethology and Behavioural Ecology*, Springer-Verlag, Berlin.

Lewontin, R.C. (1965). Selection for colonizing ability. In: *The Genetics of Colonizing Species*, Baker, H.G. & Stebbins, G.L. (Eds.), Academic Press, New York, pp. 79–94.

Lewontin, R.C. (1978). Adaptation. *Scient. Amer.* **239(3),** 212–30.

Licht, L.E. (1976). Sexual selection in toads (*Bufo americanus*). *Can. J. Zool.* **54,** 1277–84.

Lindeman, R.L. (1942). The trophic-dynamic aspects of ecology. *Ecology* **23,** 399–417.

Lindstedt, S.L. & Calder, W.A. III (1981). Body size, physiological time, and longevity of homeothermic animals. *Q. Rev. Biol.* **56,** 1–16.

Lindstedt, S.L., Miller, B.J. & Buskirk, S.W. (1986). Home range, time and body size in mammals. *Ecology* **67,** 413–8.

Llewellyn, M. & Brown, V.K. (1985). A general relationship between adult weight and the reproductive potential of aphids. *J. Anim. Ecol.* **54,** 663–73.

Lockie, J.D. (1966). Territory in small carnivores. *Symp. zool. Soc. Lond.* **18,** 143–65.

Lott, D.F. (1979). Dominance relations and breeding rate in mature male American bison. *Z. Tierpsychol.* **49,** 418–32.
Loudon, A.S.I. & Racey, P.A. (Eds.) (1987). *Reproductive Energetics in Mammals,* Academic Press, London.
MacArthur, R.H. (1972). *Geographical Ecology,* Harper & Row, New York.
MacArthur, R.H. & Wilson, E.O. (1967). *The Theory of Island Biogeography,* Princeton University Press, Princeton.
McCann, T.S. (1980). Territoriality and breeding behaviour of adult male Antarctic fur seal, *Arctocephalus gazella. J. Zool., Lond.* **192,** 295–310.
McCauley, D.E. & Wade, M.J. (1978). Female choice and the mating structure of a natural population of the soldier beetle, *Chauliognathus pennsylvanicus. Evolution* **32,** 771–5.
McGinnis, A.J. & Kasting, R. (1959). Nutrition of the pale western cutworm, *Agrotis orthogonia* Morr. (Lepidoptera: Noctuidae). I. Effects of underfeeding and artificial diets on growth and development, and a comparison of wheat sprouts of Thatcher, *Triticum aestivum* L., and Golden Ball, *T. durum* Desf., as food. *Can. J. Zool.* **37,** 259–66.
McMahon, T. (1973). Size and shape in biology. *Science* **179,** 1201–4.
McMahon, T.A. (1975). Using body size to understand the structural design of animals: quadrupedal locomotion. *J. Appl. Physiol.* **39,** 619–27.
MacMillen, R.E. & Carpenter, F.L. (1977). Daily energy cost and body weight in nectarivorous birds. *Comp. Biochem. Physiol.* **56A,** 439–41.
McNab, B.K. (1963a). Bioenergetics and the determination of home range size. *Am. Nat.* **97,** 133–40.
McNab, B.K. (1963b). A model of the energy budget of a wild mouse. *Ecology* **44,** 521–32.
McNab, B.K. (1980). Food habits, energetics, and the population biology of mammals. *Am. Nat.* **116,** 106–24.
McNeill, S. & Lawton, J.H. (1970). Animal production and respiration in animal populations. *Nature* **225,** 472–4.
Mace, G.M. (1979). *The Evolutionary Ecology of Small Mammals,* Doctor of Philosophy thesis, University of Sussex.
Mace, G.M., Harvey, P.H. & Clutton-Brock, T.H. (1983). Vertebrate home range size and energetic requirements. In: *The Ecology of Animal Movement,* Swingland, I.R. & Greenwood, P.J. (Eds.), Clarendon Press, Oxford, pp. 32–53.
Mackean, D. (1973). *Introduction to Biology,* John Murray, London.
Maiorana, V.C. (1978). An explanation of ecological and developmental constants. *Nature* **273,** 375–7.
Mansfield, A.W. (1958). The breeding behaviour and reproductive cycle of the weddell seal (*Leptonychotes weddelli* Lesson). *Falkland Islands Dependencies Survey Scientific Reports,* **No. 18,** 41 pp.
Marshall, P.T. & Hughes, G.M. (1965). *The Physiology of Mammals and other Vertebrates,* Cambridge University Press, Cambridge.
Martin, P. (1982). The energy cost of play: definition and estimation. *Anim. Behav.* **30,** 294–5.
Martin, R.D. (1981). Relative brain size and basal metabolic rate in terrestrial

vertebrates. *Nature* **293,** 57–60.
Mathavan, S. & Muthukrishnan, J. (1980). Use of larval weight and length as indices of gut content weight in some lepidopterous larvae. *Oecologia* (*Berl.*) **44,** 317–8.
Mautz, W.W. & Petrides, G.A. (1971). Food passage rate in the white-tailed deer. *J. Wildl. Mgmt.* **35,** 723–31.
May, R.M. (1979). Production and respiration in animal communities. *Nature* **282,** 443–4.
May, R.M. & MacArthur, R.H. (1972). Niche overlap as a function of environmental variability. *Proc. Nat. Acad. Sci. USA* **69,** 1109–13.
Mayer, J. (1948–1949). Gross efficiency of growth of the rat as a simple mathematical function of time. *Yale J. Biol. Med.* **21,** 415–9.
Maynard Smith, J. (1968). *Mathematical Ideas in Biology*, Cambridge University Press, Cambridge.
Maynard Smith, J. (1974). *Models in Ecology*, Cambridge University Press, Cambridge.
Maynard Smith, J. (1977). Parental investment: a prospective analysis. *Anim. Behav.* **25,** 1–9.
Maynard Smith, J. (1978). Optimization theory in evolution. *Ann. Rev. Ecol. Syst.* **9** 31–56.
Maynard Smith, J. (1980). Power and limits of optimization. In: *Evolution of Social Behaviour: Hypotheses and Empirical Tests*, Markl, H. (Ed.), Verlag Chemie, Weinheim, pp. 27–34.
Mayr, E. (1956). Geographical character gradients and climatic adaptation. *Evolution* **10,** 105–8.
Medway, W. & Kare, M.R. (1957). Water metabolism of the domestic fowl from hatching to maturity. *Am. J. Physiol.* **190,** 139–41.
Menzel, D.W. (1960). Utilization of food by a Bermuda reef fish, *Epinephelus guttatus*. *J. Conseil, Conseil Perm. Int. Exploration Mer.* **25,** 216–22.
Mertens, J.A.L. (1969). The influence of brood size on the energy metabolism and water loss of nestling great tits *Parus major major*. *Ibis* **111,** 11–16.
Millar, J.S. (1977). Adaptive features of mammalian reproduction. *Evolution* **31,** 370–86.
Millar, J.S. (1981). Pre-partum reproductive characteristics of eutherian mammals. *Evolution* **35,** 1149–63.
Mitani, J.C. & Rodman, P.S. (1979). Territoriality: the relation of ranging pattern and home range size to defendability, with an analysis of territoriality among primate species. *Behav. Ecol. Sociobiol.* **5,** 241–51.
Mitchell, B., McCowan, D. & Nicholson, I.A. (1976). Annual cycles of body weight and condition in Scottish red deer, *Cervus elaphus*. *J. Zool., Lond.* **180,** 107–27.
Mitchell, B., Staines, B.W. & Welch, D. (1977). *Ecology of Red Deer*, Institute of Terrestrial Ecology, Banchory.
Moen, A.N. (1973). *Wildlife Ecology: An Analytical Approach*, W.H. Freeman, San Francisco.
Monteith, J.L. (1973). *Principles of Environmental Physics*, Edward Arnold, London.
Moors, P.J. (1977). Studies of the metabolism, food consumption and assimi-

lation efficiency of a small carnivore, the weasel (*Mustela nivalis* L.). *Oecologia* (*Berl.*) **27,** 185–202.

Moors, P.J. (1980). Sexual dimorphism in the body size of the mustelids (Carnivora): the roles of food habits and breeding systems. *Oikos* **34,** 147–58.

Moran, N. & Hamilton, W.D. (1980). Low nutritive quality as defense against herbivores. *J. theor. Biol.* **86,** 247–54.

Morton, B. (1981). The biology and functional morphology of *Chlamydoconcha orcutti* with a discussion on the taxonomic status of the Chlamydoconchacea (Mollusca: Bivalvia). *J. Zool., Lond.* **195,** 81–121.

Morton, S.R., Hinds, D.S. & MacMillen, R.E. (1980). Cheek pouch capacity in heteromyid rodents. *Oecologia* (*Berl.*) **46,** 143–6.

Mosher, J.A. & Matray, P.F. (1974). Size dimorphism: a factor in energy savings for broad-winged hawks. *Auk* **91,** 325–41.

Mosimann, J.E. & James, F.C. (1979). New statistical methods for allometry with application to Florida red-winged blackbirds. *Evolution* **33,** 444–59.

Mueller, C.C. & Sadleir, R.M.F.S. (1977). Changes in the nutrient composition of black-tailed deer during lactation. *J. Mamm.* **58,** 421–3.

Murdie, G. (1969). The biological consequences of decreased size caused by crowding or rearing temperatures in apterae of the pea aphid, *Acyrthosiphon pisum* Harris. *Trans. R. ent. Soc. Lond.* **121,** 443–55.

Nagy, K.A. (1982). Energy requirements of free-living iguanid lizards. In: *Iguanas of the World: Their Behavior, Ecology, and Conservation*, Burghardt, G.M. & Rand, A.S. (Eds.), Noyes, Park Ridge, New Jersey, pp. 49–59.

Nagy, K.A. (1987). Field metabolic rate and food requirement scaling in mammals and birds. *Ecological Monographs* **57,** 111–28.

Nagy, K.A. & Milton, K. (1979). Energy metabolism and food consumption by wild howler monkeys (*Alouatta palliata*). *Ecology* **60,** 475–80.

Newton, I. (1979). *Population Ecology of Raptors*, T & AD Poyser, Berkhamsted.

Norton, D.W. (1970). *Thermal Regime of Nests and Bioenergetics of Chick Growth in the Dunlin* (Calidris alpina) *at Barrow, Alaska*, M.S. thesis, University of Alaska.

Nozawa, K. (1960). Distribution in the number of progeny of male parents in dairy cattle. *Jap. J. Zootech. Sci.* **31,** 188–94.

Oksanen, L., Fretwell, S.D. & Järvinen, O. (1979). Interspecific aggression and the limiting similarity of close competitors: the problem of size gaps in some community arrays. *Am. Nat.* **114,** 117–29.

Olive, C.W. (1979). *Foraging Specializations in Orb-Weaving Spiders*, Ph.D. thesis, Michigan State University.

Olive, C.W. (1981). Optimal phenology and body-size of orb-weaving spiders: foraging constraints. *Oecologia* (*Berl.*) **49,** 83–7.

Oster, G.F. & Wilson, E.O. (1978). *Caste and Ecology in the Social Insects*, Princeton University Press, Princeton.

Paloheimo, J.E. & Dickie, L.M. (1966). Food and growth of fishes. III. Relations among food, body size, and growth efficiency. *J. Fish. Res. Bd. Can.* **23,** 1209–48.

Pandian, T.J. (1967). Intake, digestion, absorption and conversion of food in the fishes *Megalops cyprinoides* and *Ophiocephalus striatus*. *Mar. Biol.* **1,** 16–32.

Parker, G.A. (1970). The reproductive behaviour and the nature of sexual selection in *Scatophaga stercoraria* L. (Diptera: Scatophagidae): IV. Epigamic recognition and competition between males for the possession of females. *Behaviour* **37,** 113–39.

Parra, R. (1978). Comparison of foregut and hindgut fermentation in herbivores. In: *The Ecology of Arboreal Folivores*, Montgomery, G.G. (Ed.), Smithsonian Institute Press, Washington, pp. 205–29.

Partridge, L. & Farquhar, M. (1981). Sexual activity reduces lifespan of male fruitflies. *Nature* **294,** 580–2.

Partridge, L., Hoffmann, A. & Jones, J.S. (1987). Male size and mating success in *Drosophila melanogaster* and *D. pseudoobscura* under field conditions. *Anim. Behav.* **35,** 468–76.

Payne, P.R. & Wheeler, E.F. (1967). Growth of the foetus. *Nature* **215,** 849–50.

Payne, P.R. & Wheeler, E.F. (1968). Comparative nutrition in pregnancy and lactation. *Proc. Nutr. Soc.* **27,** 129–38.

Pearsall, W.H. (1927). Growth studies. VI. On the relative sizes of growing plant organs. *Ann. Bot.* **41,** 549–56.

Pearson, O.P. (1948). Metabolism of small mammals, with remarks on the lower limit of mammalian size. *Science* **108,** 44.

Pearson, T.H. (1968). The feeding biology of sea-bird species breeding on the Farne Islands, Northumberland. *J. Anim. Ecol.* **37,** 521–52.

Pennycuick, C. (1972). *Animal Flight*, Edward Arnold, London.

Pentelow, F.T.K. (1939). The relations between growth and food consumption in the brown trout (*S. trutta*). *J. Exp. Biol.* **16,** 446–73.

Perrin, N., Ruedi, M. & Saiah, H. (1987). Why is the cladoceran *Simocephalus retulus* not a 'big-bang strategist'? A critique of the optimal-body-size model. *Functional Ecology* **1,** 223–8.

Peters, R.H. (1983). *The Ecological Implications of Body Size*, Cambridge University Press, Cambridge.

Petersen, B. (1950). The relation between size of mother and number of eggs and young in some spiders and its significance for the evolution of size. *Exper.* **6,** 96–8.

Phillips, A.M., Tunison, A.V., Fenn, A.H., Mitchell, C.R. & McCay, C.M. (1940). The nutrition of trout. *Cortland Hatchery Report* **9,** 32 pp.

Phillipson, J. (1960). The food consumption of different instars of *Mitopus morio* (F.) (Phalangida) under natural conditions. *J. Anim. Ecol.* **29,** 299–307.

Pianka, E.R. (1970). On *r*- and *K*-selection. *Am. Nat.* **104,** 592–7.

Pianka, E.R. (1972). *r* and *K*-selection or *b* and *d*-selection? *Am. Nat.* **106,** 581–8.

Pietsch, T.W. (1975). Precocious sexual parasitism in the deep sea ceratioid anglerfish, *Cryptopsarus cociesi* Gill. *Nature* **256,** 38–40.

Pietsch, T.W. (1979). Systematics and distribution of ceratioid anglerfish of the family Caulophrynidae with the description of a new genus and species from the Banda sea. *Contrib. Sci. Natur. Hist. Mus. Los Angeles County* **310,** 1–25.

Popp, J.L. (1978). *Male Baboons and Evolutionary Principles*, Doctor of Philosophy thesis in Biological Anthropology, Harvard University.

Potter, D.A., Wrensch, D.L. & Johnston, D.E. (1976). Aggression and mating success in male spider mites. *Science* **193,** 160–1.

Potts, D.M. (1970). Which is the weaker sex. *J. biosoc. Sci., Suppl.* **2,** 147–57.
Pough, F.H. (1980). The advantages of ectothermy for tetrapods. *Am. Nat.* **115**, 92–112.
Prange, H.D. (1977). The scaling and mechanics of arthropod exoskeletons. In: *Scale Effects in Animal Locomotion*, Pedley, T.J. (Ed.), Academic Press, London, pp. 23–36.
Prather, E.E. (1951). Efficiency of food conversion by young largemouth bass *Micropterus salmoides* (Lacépède). *Trans. Am. Fish.* **80,** 154–7.
Prothero, J.W. (1979). Maximal oxygen consumption in various animals and plants. *Comp. Biochem. Physiol.* **64A,** 463–6.
Prus, T. (1976). Experimental and field studies on ecological energetics of *Asellus aquaticus* L. (Isopoda). 2. Respiration at various temperatures as an element of energy budget. *Ekol. pol.* **24,** 604–21.
Pyke, G.H. (1978). Optimal body size in bumblebees. *Oecologia* (*Berl.*) **34,** 255–66.
Rahn, H., Paganelli, C.V. & Ar, A. (1975). Relation of avian egg weight to body weight. *Auk* **92,** 750–65.
Rajamani, M. & Job, S.V. (1976). Food utilization by *Tilapia mossambica* (Peters): a function of size. *Hydrobiologia* **50,** 71–4.
Ralls, K. (1976a). Mammals in which females are larger than males. *Q. Rev. Biol.* **51,** 245–76.
Ralls, K. (1976b). Extremes of sexual dimorphism in size in birds. *Wilson Bull.* **88,** 149–50.
Ralls, K. & Harvey, P.H. (1985). Geographic variation in size and sexual dimorphism of North American weasels. *Biol. J. Linn. Soc.* **25,** 119–67.
Randolph, P.A., Randolph, J.C. & Barlow, C.A. (1975). Age-specific energetics of the pea aphid, *Acyrthosiphon pisum*. *Ecology* **56,** 359–69.
Randolph, P.A., Randolph, J.C., Mattingly, K. & Foster, M.M. (1977). Energy costs of reproduction in the cotton rat, *Sigmodon hispidus*. *Ecology* **58,** 31–45.
Rappoldt, C. & Hogeweg, P. (1980). Niche packing and number of species. *Am. Nat.* **116,** 480–92.
Reeve, M.R. (1963). Growth efficiency in *Artemia* under laboratory conditions. *Biol. Bull.* **125,** 133–45.
Regier, H.A. (1973). Sequence of exploitation of stocks in multispecies fisheries in the Laurentian Great Lakes. *J. Fish. Res. Bd. Can.* **30,** 1992–9.
Reichle, D.E. (1968). Relation of body size to food intake, oxygen consumption, and trace element metabolism in forest floor arthropods. *Ecology* **49,** 538–42.
Reiss, M.J. (1982a). *Functional Aspects of Reproduction: Some Theoretical Considerations*, Doctor of Philosophy thesis, University of Cambridge.
Reiss, M. (1982b). Males bigger, females biggest. *New Sci.* **96,** 226–9.
Reiss, M.J. (1985). The allometry of reproduction: why larger species invest relatively less in their offspring. *J. theor. Biol.* **113,** 529–44.
Reiss, M.J. (1986a). Sexual dimorphism in body size: are larger species more dimorphic? *J. theor. Biol.* **121,** 163–72.
Reiss, M.J. (1986b). Belovsky's model of optimal moose size. *J. theor. Biol.*

122, 237–42.
Reiss, M.J. (1986c). Body size and metabolic rate: calculated exponents are independent of the units used. *J. theor. Biol.* **123,** 125–6.
Reiss, M.J. (1987a). The intraspecific relationship of parental investment to female body weight. *Functional Ecology* **1,** 105–7.
Reiss, M.J. (1987b). Why can't large animals rely on diffusion for gaseous exchange? *J. Biol. Educ.* **21,** 97–8.
Reiss, M.J. (1988). Body size and home range area. *TREE* **3,** 85–6.
Rensch, B. (1950). Die Abhängigkeit der relativen Sexualdifferenz von Körpergrösse. *Bonn. Zool. Beitr.* **1,** 58–69.
Rensch, B. (1953). [Cited, but not referenced, in Rensch, B. (1959) *loc. cit.*].
Rensch, B. (1959). *Evolution above the Species Level*, Methuen, London.
Revised Nuffield Biology (1975). *Text 2: Living Things in Action*, Longman, London.
Reynolds, R.T. (1972). Sexual dimorphism in accipiter hawks: a new hypothesis. *Condor,* **77,** 191–7.
Ricker, W.E. (1973). Linear regressions in fisheries research. *J. Fish. Res. Bd. Can.* **30,** 409–34.
Ricker, W.E. (1975). A note concerning Professor Jolicoeur's comments. *J. Fish. Res. Bd. Can.* **32,** 1294–8.
Ricker, W.E. (1979). Growth rates and models. In: *Fish Physiology, vol. VIII, Bioenergetics and Growth*, Hoar, W.S., Randall, D.J. & Brett, J.R. (Eds.), Academic Press, New York, pp. 677–743.
Ricklefs, R.E. (1968). Patterns of growth in birds. *Ibis* **110,** 419–51.
Ricklefs, R.E. (1973). Patterns of growth in birds. II. Growth rate and mode of development. *Ibis* **115,** 177–201.
Ricklefs, R.E. (1974). Energetics of reproduction in birds. In: *Avian Energetics*, Paynter, R.A. Jr. (Ed.), Publications of the Nuttall Ornithological Club, Cambridge, Massachusetts, pp. 152–292.
Ricklefs, R.E. (1977). On the evolution of reproductive strategies in birds: reproductive effort. *Am. Nat.* **111,** 458–78.
Ricklefs, R.E. (1979). Patterns of growth in birds. V. A comparative study of development in the starling, common tern, and japanese quail. *Auk* **96,** 10–30.
Ricklefs, R.E. (1982). A comment on the optimization of body size in *Drosophila* according to Roff's life history model. *Am. Nat.* **120,** 686–8.
Ricklefs, R.E., White, S.C. & Cullen, J. (1980). Energetics of postnatal growth in Leach's storm-petrel. *Auk* **97,** 566–75.
Ridley, M. (1983). *The Explanation of Organic Diversity*, Oxford University Press, Oxford.
Ridley, M. & Thompson, D.J. (1979). Size and mating in *Asellus aquaticus* (Crustacea: Isopoda). *Z. Tierpsychol.* **51,** 380–97.
Roberts, D.F. (1977). Assortative mating in man. *Bull. Eugenics Soc. Suppl.* 2.
Roberts, M.B.V. (1986a). *Biology for Life*, 2nd edn, Nelson, Walton-on-Thames, Surrey.
Roberts, M.B.V. (1986b). *Biology: A Functional Approach*, 4th edn, Nelson, Walton-on-Thames, Surrey.

Robertson, F.W. (1960). The ecological genetics of growth in *Drosophila*. 1. Body size and development time on different diets. *Genet. Res., Camb.* **1,** 288–304.

Robinson, M.H. & Robinson, B. (1979). By dawn's early light: matutinal mating and sex attractants in a Neotropical mantid. *Science* **205,** 825–7.

Robinson, S.K. (1986). Benefits, costs, and determinants of dominance in a polygynous oriole. *Anim. Behav.* **34,** 241–55.

Roff, D. (1981). On being the right size. *Am. Nat.* **118,** 405–22.

Roff, D.A. (1983). An allocation model of growth and reproduction in fish. *Can. J. Fish. Aquat. Sci.* **40,** 1395–404.

Roth, V.L. (1981). Constancy in the size ratios of sympatric species. *Am. Nat.* **118,** 394–404.

Rowlinson, P. & Jenkins, M. (1982). *Human Biology: An Activity Approach*, Cambridge University Press, Cambridge.

Rudder, B.C.C. (1979). *The Allometry of Primate Reproductive Parameters*, Ph.D., University College, London.

Russell, E.M. (1982). Patterns of parental care and parental investment in marsupials. *Biol. Rev.* **57,** 423–86.

Ryan, M.S. (1983). Sexual selection and communication in a neotropical frog, *Physalaemus pustulosus*. *Evolution* **37,** 261–72.

Sadleir, R.M.F.S. (1980). Milk yield of black-tailed deer. *J. Wildl. Mgmt.* **44,** 472–8.

Sadleir, R.M.F.S., Casperson, K.D. & Harling, J. (1973). Intake and requirements of energy and protein for the breeding of wild deermice, *Peromyscus maniculatus*. *J. Reprod. Fertil., Suppl.* **19,** 237–52.

Sarrus & Rameaux (1839). Mémoire adressé à l'Académie Royale. *Bulletin de l'académie royale de médicine* **3,** 1094–100.

Scheffer, V.B. & Wilke, F. (1953). Relative growth in the northern fur seal. *Growth* **17,** 129–45.

Schiemer, F., Duncan, A. & Klekowski, R.Z. (1980). A bioenergetic study of a benthic nematode, *Plectus palustris* de Man 1880, throughout its life cycle. II. Growth, fecundity and energy budgets at different densities of bacterial food and general ecological considerations. *Oecologia* (*Berl.*) **44,** 205–12.

Schladweiler, P. & Stevens, D.R. (1973). Weights of moose in Montana. *J. Mamm.* **54,** 772–5.

Schmidt-Nielsen, K. (1972). Locomotion: energy cost of swimming, flying and running. *Science* **177,** 222–8.

Schmidt-Nielsen, K. (1983). *Animal Physiology: Adaptation and Environment*, 3rd edn, Cambridge University Press, Cambridge.

Schmidt-Nielsen, K. (1984). *Scaling: Why is Animal Size so Important?*, Cambridge University Press, Cambridge.

Schoener, T.W. (1968). Sizes of feeding territories among birds. *Ecology* **49,** 123–41.

Schoener, T.W. (1969). Models of optimal size for solitary predators. *Am. Nat.* **103,** 277–313.

Schoener, T.W. & Schoener, A. (1978). Inverse relation of survival of lizards with island size and avifaunal richness. *Nature* **274,** 685–7.

Searcy, W.A. (1979). Sexual selection and body size in male red-winged blackbirds. *Evolution* **33,** 649–61.

Searcy, W.A. (1980). Optimum body sizes at different ambient temperatures: an energetics explanation of Bergmann's rule. *J. theor. Biol.* **83,** 579–93.

Sebens, K.P. (1979). The energetics of asexual reproduction and colony formation in benthic marine invertebrates. *Amer. Zool.* **19,** 683–97.

Selander, R.K. (1972). Sexual selection and dimorphism in birds. In: *Sexual Selection and the Descent of Man 1871–1971*, Campbell, B. (Ed.), Aldine Publishing Company, Chicago, pp. 180–230.

Short, H.L. (1964). Postnatal stomach development of white-tailed deer. *J. Wildl. Mgmt.* **28,** 445–58.

Shubin, I.G. & Shubin, N.G. (1975) [Sexual dimorphism and its peculiarities in mustelines (Mustelidae, Carnivora).] [In Russian] *Z. Obshch. Biol.* **36,** 283–90.

Shuster, S.M. (1981). Sexual selection in the Socorro isopod *Thermosphaeroma thermophilum* (Cole) (Crustacea: Peracarida). *Anim. Behav.* **29,** 698–707.

Sibly, R.M. & Calow, P. (1986). *Physiological Ecology of Animals: An Evolutionary Approach*, Blackwell Scientific Publications, Oxford.

Simberloff, D. & Boecklen, W. (1981). Santa Rosalia reconsidered: size ratios and competition. *Evolution* **35,** 1206–28.

Simpson, A.M., Webster, A.J.F., Smith, J.S. & Simpson, C.A. (1978). The efficiency of utilization of dietary energy for growth in sheep (*Ovis ovis*) and red deer (*Cervus elaphus*). *Comp. Biochem. Physiol.* **59A,** 95–9.

Sinclair, A.R.E. (1977). *The African Buffalo: A Study of Resource Limitation of Populations*, University of Chicago Press, Chicago.

Smith, C.C. & Fretwell, S.D. (1974). The optimal balance between size and number of offspring. *Am. Nat.* **108,** 499–506.

Smith, K.L. Jr. (1973). Energy transformation by the sargassum fish *Histrio histrio* (L.). *J. exp. mar. Biol. Ecol.* **12,** 219–27.

Smith, R.J. (1980). Rethinking allometry. *J. theor. Biol.* **87,** 97–111.

Sokal, R.R. & Rohlf, F.J. (1969). *Biometry: The Principles and Practice of Statistics in Biological Research*, W.H. Freeman, San Francisco.

Soper, R. & Tyrell Smith, S. (1979). *Modern Biology for First Examinations*, Macmillan Education, Basingstoke.

Southwood, T.R.E. (1976). Bionomic strategies and population parameters. In: *Theoretical Ecology: Principles and Applications*, May, R.M. (Ed.), Blackwell Scientific Publications, Oxford, pp. 26–48.

Spight, T.M. & Emlen, J. (1976). Clutch sizes of two marine snails with a changing food supply. *Ecology* **57,** 1162–78.

Stahl, W.R. (1962). Similarity and dimensional methods in biology. *Science* **137,** 205–12.

Stamps, J.A. (1983). Sexual selection, sexual dimorphism and territoriality in lizards. In: *Lizard Ecology: Studies of a Model Organism*, Huey, R.B., Pianka, E.R. & Schoener, T.W. (Eds.), Harvard University Press, Cambridge, Massachusetts, pp. 169–204.

Staples, D.J. & Nomura, M. (1976). Influence of body size and food ration on the energy budget of rainbow trout *Salmo gairdneri* Richardson. *J. Fish Biol.* **9,** 29–43.

Stearns, S.C. (1976). Life-history tactics: a review of the ideas. *Q. Rev. Biol.* **51,** 1–47.

Stearns, S.C. (1980). A new view of life-history evolution. *Oikos* **35,** 266–81.

Steel, E.A. (1961). Some observations on the life history of *Asellus aquaticus* (L.) and *Asellus meridianus* Racovitza (Crustacea: Isopoda). *Proc. zool. Soc. Lond.* **137,** 71–87.

Strong, K.W. & Daborn, G.R. (1979). Growth and energy utilisation of the intertidal isopod *Idotea baltica* (Pallas) (Crustacea: Isopoda). *J. exp. mar. Biol. Ecol.* **41,** 101–23.

Strong, K.M. & Daborn, G.R. (1980). The influence of moulting on the ingestion rate of an Isopod crustacean. *Oikos* **34,** 159–62.

Sullivan, B.K. (1982). Sexual selection in Woodhouse's toad (*Bufo woodhousei*). I. Chorus organization. *Anim. Behav.* **30,** 680–6.

Sullivan, B.K. (1987). Sexual selection in Woodhouse's toad (*Bufo woodhousei*). III. Seasonal variation in male mating success. *Anim. Behav.* **35,** 912–9.

Sushchenya, L.M. & Khmeleva, N.N. (1967). Consumption of food as a function of body weight in crustaceans. *Dokl. Acad. Sci. USSR* (*English*) **176,** 559–62.

Swihart, R.K., Slade N.A. & Bergstrom, B.J. (1988). Relating body size to the rate of home range use in mammals. *Ecology* **69,** 393–9.

Tanner, J.M. (1978). *Foetus into Man: Physical Growth from Conception to Maturity*, Open Books, London.

Taylor, C.R. (1977). The energetics of terrestrial locomotion and body size in vertebrates. In: *Scale Effects in Animal Locomotion*, Pedley, T.J. (Ed.), Academic Press, London, pp. 127–41.

Taylor, C.R. (1980). Evolution of mammalian homeothermy: a two-step process? In: *Comparative Physiology: Primitive Mammals*, Schmidt-Nielsen, K., Bolis, L. & Taylor, C.R. (Eds.), Cambridge University Press, Cambridge, pp. 100–11.

Taylor, C.R., Maloiy, G.M.O., Weibel, E.R., Langman, V.A., Kamau, J.M.Z., Seeherman, H.J. & Heglund, N.C. (1980). Design of the mammalian respiratory system. III. Scaling maximum aerobic capacity to body mass: wild and domestic mammals. *Respir. Physiol.* **44,** 25–37.

Taylor, L.R. (1975). Longevity, fecundity and size: control of reproductive potential in a polymorphic migrant *Aphis fabae* Scop. *J. Anim. Ecol.* **44,** 135–63.

Taylor, St C.S. (1965). A relation between mature weight and time taken to mature in mammals. *Anim. Prod.* **7,** 203–20.

Taylor, St C.S. (1968). Time taken to mature in relation to mature weight for sexes, strains, and species of domesticated mammals and birds. *Anim. Prod.* **10,** 157–69.

Taylor, W.D. & Shuter, B.J. (1981). Body size, genome size, and intrinsic rate of increase in ciliated Protozoa. *Am. Nat.* **118,** 160–72.

Thompson, D'A.W. (1942). *On Growth and Form*, Cambridge University Press, Cambridge.

Thurling, D.J. (1980). Metabolic rate and life stage of the mites *Tetranychus cinnabarinus* Boisd. (Prostigmata) and *Phytoseiulus persimilis* A–H. (Mesostigmata). *Oecologia* (*Berl.*) **46,** 391–6.

Tracy, C.R. (1977). Minimum size of mammalian homeotherms: role of the thermal environment. *Science* **198,** 1034–5.

Trivers, R.L. (1972). Parental investment and sexual selection. In: *Sexual Selection and the Descent of Man 1871–1971*, Campbell, B. (Ed.), Aldine Publishing Company, Chicago, pp. 136–79.

Trivers, R.L. (1976). Sexual selection and resource-accruing abilities in *Anolis garmani*. *Evolution* **30,** 253–69.

Tucker, V.A. (1977). Scaling and avian flight. In: *Scale Effects in Animal Locomotion*, Pedley, T.J. (Ed.), Academic Press, London, pp. 497–509.

Tunison, A.V., Phillips, A.M., McCay, C.M., Mitchell, C.R. & Rodgers, E.O. (1939). The nutrition of trout. *Cortland Hatchery Report* No. 8, New York Conservation Department, Albany, New York.

Tuomi, J. (1980). Mammalian reproductive strategies: a generalised relation of litter size to body size. *Oecologia* (*Berl.*) **45,** 39–44.

Tuomi, J., Hakala, T. & Haukioja, E. (1983). Alternative concepts of reproductive effort. Costs of reproduction and selection in life-history evolution. *Amer. Zool.* **23,** 25–34.

Turner, F.B. (1970). The ecological efficiency of consumer populations. *Ecology* **51,** 741–2.

Turrell, F.M. (1961). Growth of the photosynthetic area of citrus. *Bot. Gaz.* **122,** 284–98.

Uetz, G.W. (1977). Coexistence in a guild of wandering spiders. *J. Anim. Ecol.* **46,** 531–41.

Ursin, E. (1967). A mathematical model of some aspects of fish growth, respiration and mortality. *J. Fish Res. Bd. Can.* **24,** 2355–453.

Ursin, E. (1979). Principles of growth in fishes. *Symp. zool. Soc. Lond.* **44,** 63–87.

Van der Drift, J. (1951). Analysis of the animal community in a beech forest floor. *Tijdschrift voor entomologie* **94,** 1–168.

Van Devender, R.W. (1978). Growth ecology of a tropical lizard, *Basiliscus basiliscus*. *Ecology* **59,** 1031–8.

Van Hook, R.I., Nielsen, M.G. & Shugart, H.H. (1980). Energy and nitrogen relations for a *Macrosiphum liriodendri* (Homoptera: Aphididae) population in an east Tennessee *Liriodendron tulipifera* stand. *Ecology* **61,** 960–75.

Venus, J.C. & Causton, D.R. (1979). Plant growth analysis: the use of the Richards Function as an alternative to polynomial exponentials. *Ann. Bot.* **43,** 623–32.

Veuille, M. & Rouault, J. (1980). Experimental evidence of sexual selection based on male body size in *Jaera* (Isopoda: Asellota). *Exper.* **36,** 549–50.

von Bertalanffy, L. (1960). Principles and theory of growth. In: *Fundamental Aspects of Normal and Malignant Growth*, Nowinski. W.W. (Ed.), Elsevier, Amsterdam, pp. 137–259.

Wade, M.J. & Arnold, S.J. (1980). The intensity of sexual selection in relation to male sexual behaviour, female choice, and sperm precedence. *Anim. Behav.* **28,** 446–61.

Waldbauer, G.P. (1968). The consumption and utilization of food by insects. *Adv. Insect Physiol.* **5,** 229–88.

Walsberg, G.E. (1983a). Ecological energetics: what are the questions? In: *Perspectives in Ornithology*, Bush, A.H. & Clark, G.A. Jr. (Eds.), Cambridge University Press, Cambridge, pp. 135–58.

Walsberg, G.E. (1983b). Avian ecological energetics. In: *Avian Biology Volume VII*, Farner, D.S., King, J.R. & Parkes, K.C. (Eds.), Academic Press, New York, pp. 161–220.

Walter, H. (1979). *Eleonora's Falcon: Adaptations to Prey and Habitat in a Social Raptor*, University of Chicago Press, Chicago.

Ware, D.M. (1980). Bioenergetics of stock and recruitment. *Can. J. Fish. Aquat. Sci.* **37,** 1012–24.

Waters, T.F. (1969). The turnover ratio in production ecology of freshwater invertebrates. *Am. Nat.* **103,** 173–85.

Watson, A. (Ed.) (1970). *Animal Populations in Relation to their Food Resources*, Blackwell Scientific Publications, Oxford.

Watt, A.D. (1979). The effect of cereal growth stages on the reproductive activity of *Sitobion avenae* and *Metopolophium dirhodum*. *Ann. appl. Biol.* **91,** 147–57.

Way, M.J. (1968). Intra-specific mechanisms with special reference to aphid populations. *Symp. R. ent. Soc. Lond.* **4,** 18–36.

Wellings, P.W., Leather, S.R. & Dixon, A.F.G. (1980). Seasonal variation in reproductive potential: a programmed feature of aphid life cycles. *J. Anim. Ecol.* **49,** 975–85.

Wells, K.D. (1977). The social behavior of anuran amphibians. *Anim. Behav.* **25,** 666–93.

Wells, K.D. (1979). Reproductive behavior and male mating success in a neotropical toad, *Bufo typhonius*. *Biotropica* **11,** 301–7.

Western, D. (1979). Size, life history and ecology in mammals. *Afr. J. Ecol.* **17,** 185–204.

Wheeler, P. & Greenwood, P.J. (1983). The evolution of reversed sexual dimorphism in birds of prey. *Oikos* **40,** 145–9.

White, T.C.R. (1978). The importance of a relative shortage of food in animal ecology. *Oecologia* (*Berl.*) **33,** 71–86.

Whitford, W.G. & Hutchinson, V.H. (1967). Body size and metabolic rate in salamanders. *Physiol. Zool.* **40,** 127–33.

Whittaker, R.H. & Woodwell, G.M. (1968). Dimension and production relations of trees and shrubs in the Brookhaven forest, New York. *J. Ecol.* **56,** 1–25.

Wiegert, R.G. (1961). Respiratory energy loss and activity patterns in the meadow vole, *Microtus pennsylvanicus pennsylvanicus*. *Ecology* **42,** 245–53.

Wiens, J.A. & Rotenberry, J.T. (1981). Morphological size ratios and competition in ecological communities. *Am. Nat.* **117,** 592–9.

Wieser, W. (1985). A new look at energy conversion in ectothermic and endothermic animals. *Oecologia* (*Berl.*) **66,** 506–10.

Wiley, R.H. (1974). Evolution of social organization and life-history patterns among grouse. *Q. Rev. Biol.* **49,** 201–27.

Wilkie, D.R. (1977). Metabolism and body size. In: *Scale Effects in Animal Locomotion*, Pedley, T.J. (Ed.), Academic Press, London, pp. 23–36.

Williams, G.C. (1959). Ovary weight of darters: a test of the alleged association of parental care with reduced fecundity in fishes. *Copeia* **1959,** 18–24.
Williams, G.C. (1966). Natural selection, the costs of reproduction and a refinement of Lack's principle. *Am. Nat.* **100,** 687–90.
Willner, L.A. & Martin, R.D. (1985). Some basic principles of mammalian sexual dimorphism. In: *Human Sexual Dimorphism*, Ghesquiere, J., Martin, R.D. & Newcombe, F. (Eds.), Taylor & Francis, London, pp. 1–42.
Wilson, E.O. (1975). *Sociobiology: The New Synthesis*, Belknap Press, Cambridge, Massachusetts.
Wilson, E.O. (1980). Caste and division of labor in leaf-cutter ants (Hymenoptera: Formicidae: *Atta*). II. The ergonomic optimization of leaf cutting. *Behav. Ecol. Sociobiol.* **7,** 157–65.
Woodward, B.D. (1982). Male persistence and mating success in Woodhouse's toad (*Bufo woodhousei*). *Ecology* **63,** 583–5.
Wootton, R.J. (1979). Energy costs of egg production and environmental determinants of fecundity in teleost fishes. *Symp. zool. Soc. Lond.* **44,** 133–59.
Wrangham, R.W. & Smuts, B.B. (1980). Sex differences in the behavioural ecology of chimpanzees in the Gombe National Park, Tanzania. *J. Reprod. Fert., Suppl.* **28,** 13–31.
Wratten, S.D. (1977). Reproductive strategy of winged and wingless morphs of the aphids *Sitobion avenae* and *Metopolophium dirhodum*. *Ann. appl. Biol.* **85,** 319–31.
Young, S.R. (1979). Effect of temperature change on the metabolic rate of an Antarctic mite. *J. Comp. Physiol.* **131,** 341–6.
Zar, J.H. (1968). Calculation and miscalculation of the allometric equation as a model in biological data. *BioScience* **18,** 1118–20.
Zeuthen, E. (1953). Oxygen uptake as related to body size in organisms. *Q. Rev. Biol.* **28,** 1–12.

Index

The following terms used throughout the text are not included in the index: allometry, body weight, confidence limits, correlation, female, length, male, model, quantitative, regression, reproduction, size.

Acanthops, 106
Acyrthosiphon, 34, 52
adult motor patterns, 71
Aepyornis, 77
age
  at end of parental investment, 44
  at first reproduction, 47, 126
  at fledgling, 45–6, 72
  at maturity, 43–4, 72
  at weaning, 46
Agelaius, 97
aggression, 92
Agrotis, 52
Alaskozetes, 12
Alces, 97
allometric equation, 1
Allouatta, 9
alveoli, 131
ambient temperature, 16
amino acids, 97
anabolism, 40
animal
  activity rate, 84
  density, 69
  structures, 62
animal's temperature, *see* body temperature
annual parental investment, 90
Anolis, 93–4, 103–4, 108
antler weight, 95, 140–1
Aphis, 34
arboreality, 120–1, 123
Arctocephalus, 114
Ardeotis, 87
Artemia, 18, 52
arteries, 77
Arvicola, 9
Asellus, 12, 32–3, 37, 109
assimilation
  efficiency, 58
  assortative mating for size, 95, 105
Atta, 13, 85, 90
average daily metabolic rate, 7–16, 59, 66, 69, 79, 101, 114, 130

barrier thickness, 131
basal metabolic rate, 8, 10–1, 14, 59, 65–6, 76–9, 130–1
basal metabolism, *see* basal metabolic rate
basal rate of energy expenditure, *see* basal metabolic rate
Basiliscus, 108
behaviour, 143
benefits of play, *see* play
Bergmann's rule, 135–7
bioenergetic advantage, 91
biological cycles, 60
biological time, 47, 67
bipedality, 85
birth, 27, 142
Bison, 93–4, 103–4
bite size, 21

blood, 131
  volume, 81
body reserves, 136
body temperature, 16, 136
Bombus, 84
Bombyx, 52
bone lengths, 11
Bonellia, 108
brain
  mass, 4, 140
  size, *see* brain mass
  weight, *see* brain mass
Branchinecta, 12
breeding groups, 120
breeding season, 61, 94, 106–7
Brevicoryne, 34
Bufo, 93–4, 103–6, 123
Bursera, 139

Calidris, 52
caloric content of food, 78–9
caloric requirements, *see* energy requirements
canines, 141
capillaries, 131
carapace length, 122
carcase weight, 95
cardiac cycle, 47
cardiac output, 88
carrying capacity, 68–9, 73
catabolism, 40
Cervus, 78, 81–3, 93–5, 103–4, 108
Chauliognathus, 105
Choristoneura, 52
chorus, 106
Chossat's Law, 136–7
chronological time, 67
cilia, 136, 138
Cimex, 52
circulatory system, 131–2, 134, 137
Clethrionomys, 8–9
climate, 135
clutch
  number, 25–6
  size, 25, 68, 142
  volume, 25–6
  weight, 25–7
coefficient of variation, 62
Coenagrion, 112
coexistence, 62–4, 72
colony, 84–5
commercial value, 140
community, 126
competition, 120, 123–4, 126
competitive ability, 124
conception, 27
conduction, 136
conflict, *see* competition
consort success, 94
conspecifics, 66–7
constraint equations, 88, 90
contraction cycle, 88
contraction time of the heart, 88
convection, 16, 136
coordination, 71, 140
copulation, 92, 94
Corvus, 77
cost
  of movement, *see* locomotion
  of play, *see* play
Cricetus, 9
cumulative parental investment, 27–8
cycle lengths, 47, 88
Cyclops, 13
Cygnus, 87
Cylindrojulus, 17
Cyprinodon, 52

daily energetic needs, *see* daily energy expenditure
daily energy budget, 7
daily energy expenditure, 7, 65, 67, 69–71
daily metabolic requirements, *see* daily energy expenditure
Daphnia, 18
death, 138
deferred reproduction, 124
dehydration, 137
delayed reproduction, *see* deferred reproduction
Delichon, 52
Dendrocygna, 8, 52
dependent variable, 3
desiccation, 138
determinate growth, 53
development time, 86
diffusion, 129–34, 137
digestibility, 78–80
digestive enzymes, 20
digestive surfaces, 137
digestive system, 80
dimensional analysis, 47
dimorphism, *see* sexual dimorphism
dominance, 92, 95
Drepanosiphum, 34
Drosophila, 36, 86–7, 110

ecological cycles, 61
ecology, 25, 128, 143
ecosystem, 58
egg laying, 27
egg production, 61
egg weight, 44, 63

elastic similarity, 11
Eleutherodactylus, 105
Enallagma, 112
enamel ridge folding, 134
encephalization quotients, 140
energetic efficiency, 124
energetic investment, *see* parental investment
energetic needs, *see* energy requirements
energy
  absorption, *see* energy intake
  assimilation, *see* energy intake
  available for reproduction, *see* parental investment
  budgets, 48–62, 71
  control, 63
  cost of play, *see* play
  devoted to reproduction, *see* parental investment
  expenditure, *see* energy requirements
  intake, 7, 17–21, 23–5, 29–31, 36, 54, 57–62, 69, 71–2, 75, 78–80, 84, 86–7, 89–90, 96–101, 107, 114, 120, 126–7, 134–5
  invested in reproduction, *see* parental investment
  requirements, 7, 23–5, 27, 29–31, 33, 36, 60, 63, 67, 72, 77, 87–9, 96–100, 120, 126–7, 135, 142
  reserves, 143
environment, 60
environmental fluctuations, 63
Epinephelus, 52
epithelium, 130
Eucallipterus, 34
evaporation, 16
evolution, 143
evolutionary time, 63
existence energy requirements, *see* existence metabolism
existence metabolism, 76, 124
exploitation efficiency, 58
explosive breeding, 113

farming methods, 54, 61–2
fasting, 77
fecundity, 32–6, 68, 71
feeding
  area, *see* home range size
  behaviour, 25
  divergence, 126
  energy expenditure on, 76
  time, 69, 76, 99
fermentation rates, 82, 84
field metabolic rate, 7
fights, 105, 110
fitness, 70–1
fledgling weight, 45
flight, 16, 30, 87–8
folding of the intestines, 135
food
  bulkiness, 78–9
  consumption, *see* energy intake
  distribution, 128
  gathering structures, 135, 137
  intake, *see* energy intake
  output, 62
  requirements, *see* energy intake
  web, 62
  weight, 82
foraging, 91
functional tissue, 54

Gadus, 52–3
gaseous exchange, 129, 132
Gause's competitive exclusion principle, 64
generation time, 44, 46–7, 68, 72
genetics, 143
geometric similarity, 11, 21, 134
gestation, 44–5, 48, 67, 72
gills, 129, 132
Glaucomys, 8–9
Glomeris, 17
glucose, 140
Gompertz equation, 39
gravity, 85
gravitational fields, *see* gravity
gravitational load, 14
grinding surface area, 134
growth, 10, 24, 31–2, 39–45, 48–54, 56, 58, 60–1, 70–2, 78, 86, 96, 108, 124, 139, 141
growth efficiency, 48–56, 61, 71–2, 142
Gymnogyps, 87

habitat productivity, 66
half-life of drugs, 47
Halichoerus, 106
harems, 110, 121
hatching, 46, 94
heart
  beat, 88
  rate, 88
  size, 88
heat gain, 136, 138
heat loss, 16, 135–6, 138
hibernation, 61
high-energy foods, 66
Histrio, 52
home range area, *see* home range size
home range size, 64–9, 72
homeothermy, 58–61, 72, 109, 131
Homo, 105
hovering, 30, 87–8

husbandry, 61
Hyla, 105
hypsodonty, 134–5, 137

Idotea, 12, 18, 48
incisor breadth, 20
incubation time, 44–5, 47, 72
independent variable, 3
indeterminate growth, 53
injury, 69
insulating areas, 136
intelligence, 140
intermale conflict, 91
isometry, 1, 134
iteroparity, 48

juvenile growth rate, 43

K, *see* carrying capacity
K selection, 64
Kareius, 52
Kinosternon, 122

lactation, 27–8, 45–7, 67, 72
latitude, 135–6
Law of Trophic Efficiency, 57
Lepomis, 52
Leptonychotes, 110
Lestes, 52
Libellula, 112
life cycle, 57, 60, 72
lifespan, 28, 33, 36, 46–7, 60–1, 67–8, 89–90, 109, 134
lifetime
  egg production, 112–13
  energy expenditure, 28
  energy intake, 28
  parental investment, 89–90
  reproductive success, 60, 89, 96, 109, 112–13
Limanda, 52, 80
litter
  size, 25, 42–3, 142
  mass, *see* litter weigth
  weaning weight, 25–7
  weight, 25–8, 45, 48, 72
livestock, 61
locomotion, 136, 138
  cost of, 16, 78–9, 110
locomotor ability, 92
Logistic equation, 39, 41
longevity, *see* lifespan
lophophores, 135
lungs, 129, 132
Lygaeus, 52–3

Macrosiphum, 12
maintenance costs, *see* metabolic rate
mating, 94
  success, *see* reproductive success
maturity, size at, 124
maximal rates
  of aerobic energy expenditure, *see* maximal rates of oxygen consumption
  of oxygen consumption, 2–3, 130–1
maximum heat production, 76
mean animal activity rate, *see* animal activity rate
Megalops, 17–18, 50, 52, 55, 80
meat production, 61
memory, 140
Mendel's laws, 143
Menippe, 18
Mesocyclops, 12
metabolic
  activity, 46
  rate, 8, 12–13, 20, 43, 46, 49, 53–4, 57–60, 63, 65–9, 75–80, 126, 130–4, 136–8, 140
  power, 61
  requirements, *see* metabolic rate
Metopolophium, 34–5
microcosm, 64
Micromys, 9
Micropterus, 52
Microtus, 9
milk, 41
  production, 29
  yield, 26–7
Mirounga, 108
Mitopus, 17
mitotic tissue, 54
mobility, 110
molar surface grinding area, 134
molarization, 134
monogamy, 98, 121, 126
mortality, 91
mouth volume, 21
muscle, 140
  contraction time, 47
Mustela, 9, 108, 122
Myzus, 33, 35

natural rate of increase, 67–9, 73, 86–7
natural selection, 60, 72, 141
nectar, 20, 84
neonate weight, *see* offspring, weight of
net daily energy expenditure, *see* daily energy expenditure
net energy intake, *see* energy intake
niche, 62, 64, 124
nitrogen requirements, 98
Notodiaptomus, 12
nutrient transport, 77

nutrients, 97, 126
Oceanodroma, 51–2, 55
Odocoileus, 46, 80
oestrus, 107
offspring
  fitness, 142
  number of, 63, 68
  weight, 25, 27, 44–5, 47, 63, 142
Oncopeltus, 52
Oncorhynchus, 10
Ophiocephalus, 17–18, 50, 52, 55
optimization, 5, 85, 88, 90
Orchestia, 18
organ size, 140–1
organic matter, 20
overexploitation, 62
overfishing, 62
oviposition sites, 105
oxygen
  consumption, 2–3
  requirements, *see* metabolic rate
  transport, 77

Pan, 107
Papio, 93–4, 103–4, 142
parental investment, 23–37, 41–4, 47, 51–4, 63, 70–1, 79, 89, 91, 97–102, 112–13, 115, 134, 141–3
Paropsis, 52
partial pressure, 133–4
Passer, 8, 52
patches, 66
Pelacanus, 87
permeability constant, 133–4
Phonoctonus, 52
Physalaemus, 109
physiological relevance, 42–3
physiology, 25, 143
Phytoseiulus, 12
placental surfaces, 29
plants, 139–40
play, 69–71, 73
Plectus, 12, 18, 48, 52
Pleuronectus, 52
poikilothermy, 58–61, 72, 109, 131
polygyny, 105, 110, 117, 120, 123, 127–8, 140
population
  density, 62–3
  genetics, 69, 143
  increase, 87
  number, *see* population size
  size, 64
post-spawning survival, 33
pouch life, 46
predation, 69, 91, 109, 125, 143
premolars, 134
prenatal growth rate, 45
probability tests, 5–6
Procyon, 46
production, 46, 56–60, 62, 68–9, 72
production efficiency, 58, 60, 72
productivity, *see* production
productivity/biomass ratios, 54, 56–7, 72
promiscuity, 110
protection, 132
protein, 97–8
Pseudopleuronectus, 52

Quiscalus, 91

r, *see* natural rate of increase
r selection, 64, 73, 127
radiation, 16, 136
Rana, 93–4, 102–4
rank, 96
rate of reproduction, 62, 72, 109
regeneration of food supply, 67, 69
relative consort success, *see* consort success
relative growth rate, 41, 43–4, 71
reproductive
  efficiency, 127
  effort, *see* parental investment
  growth rate, 45, 48, 72
  rate, 46
  success, 91–7, 99–110, 112, 114–15, 140, 143
resource axis, 125
resource food continuum, 62
respiration, 56, 58
respiratory
  gases, 130
  structures, 130–2, 134
  surfaces, 135, 137
resting, 69, 71
  metabolism, 14, 69
  oxygen consumption, *see* resting metabolism
retention time, 80
reversed sexual dimorphism, 118–119
Rhopalosiphum, 35
rumen, 78–80, 82
ruminant, *see* rumen
running costs, 16

Salmo, 10, 18, 52–3
Salvelinus, 52
saturated environments, 124
Scatophaga, 105
scramble competition, 107
seasonal environments, 143
secondary sex characters, 124
seed crop, 140

selection, 87, 124, 134, 141–2
  intrasexual, 117
  sexual, 117, 124, 127, 141
selective coefficients, 69–71
selective forces, *see* selection
semelparity, 29, 36, 48
sensitivity analysis, 86, 88, 90
sex hormones, 124
sexual dimorphism, 91–128, 143
Sitobion, 35
skin, 132
social classifications, 123
social groups, 127
social organization, 127
socionomic sex ratio, 120–1, 127
sodium, 97
Sorex, 8–9
spawning, 94
specialization, 140
species
  coexistence, *see* coexistence
  energy control, *see* energy control
  survival, 63
specific metabolic power, 88
speed, 84
sperm
  competition, 141
  production, 110
Stalia, 52
standard metabolic rate, 10, 59, 77
standard metabolism, *see* standard metabolic rate
standing crop, 57, 72
starvation, 109, 136–8
Sterna, 52
Sternotherus, 122
stomach, 80, 82–3
stored reserves, 61
strength, 71
stroke volume, 88
Struthio, 77
sulphur requirements, 98
surface
  area, 16, 129–31, 134, 136–7
  area/volume ratio, 126, 129–38
  law, 14
survival, 61, 71, 127
survivorship, 28, 69
swimming, 30

Tamiasciurus, 8–9
teeth, 134–5, 138
temperature regulation, 76
terrestriality, 120
territory, 122
  defence of, 92, 112
  size, 65, 67
testes size, 141
Tetranychus, 12
thermal conductance, 16
Thermocyclops, 13
thermoneutrality, 10, 77
thermoregulation, 79
Thermosphaeroma, 105
throughput
  rate, 80
  time, 80
Tilapia, 52–3
time
  spent assimilating energy, 98–101, 109, 112, 142
  spent locomoting, 110
  spent trying to reproduce, 98–101, 106–7, 109, 111–15
  to metabolize fat stores, 47
  to reach adult size, 67
  to reach independence, 67
  to reach reproductive maturity, 143
  to starve, 138
tissues, 132, 137
tooth area, 134
total daily expenditure, 65
total energetic requirements, 67
total metabolic expenditure, 33
total metabolic power, 60, 87
trophic
  apparatus, 62
  efficiency, 57–8
  level, 58, 65, 73
turnover ratio, 57, 72

Uca, 105

Van der Drift constant, 17
variance
  in reproductive success, 106–7
  in the number of mates, 141
viscous losses, 77
volume law, 131
von Bertalanffy equation, 39–41

weaning weight, 27, 45
web area, 86
weight gain, *see* growth
wing length, 125

zonation classes, 123